Robin Dhrushka
2008

BARREL RACING
Training the Wright Way

ED AND MARTHA WRIGHT
with Glory Ann Kurtz

BARREL RACING
Training the Wright Way

ED AND MARTHA WRIGHT
with Glory Ann Kurtz

Kurtz and Wright

Dublin, Texas

BARREL
RACING
Training the Wright Way

Authors	Ed and Martha Wright
Editor	Glory Ann Kurtz
Photographers	Glory Ann Kurtz
	Kathy Swan
	Rick Swan
	Kenneth Springer
Graphic Designer	Rob Feinberg
Illustrator	Rob Feinberg
Production Asst.	Victoria J. Allen

Copyright © 1999

All rights reserved. No part of this book may be reproduced without written permission, except by a reviewer who may quote brief passages or reproduce illustrations in a review with appropriate credits. Nor may any part of this book be reproduced by any means graphically, electronically, or mechanically, including photocopying, recording, taping, or information storage and retrieval system without written permission from the publisher.

Library of Congress Cataloging-in-Publication Data

Wright, Ed, 1946-
    Barrel racing : training the Wright way / Ed and Martha Wright with Glory Ann Kurtz ; [edited by Glory Ann Kurtz].
    p. cm.
    Includes index.
    ISBN 0-9625898-9-6
    ISBN 0-9720477-0-0
    1. Barrel racing. 2. Horses—Training. I. Wright, Martha, 1951- . II. Kurtz, Glory Ann, 1939- . III. Title.
GV1834.45.B35W75   1999
791.8'4—dc21   99-17227
                                                            CIP

Published by
Kurtz and Wright
312 Private Rd. 1184
Dublin, Texas 76446
Tel: 254-968-3661

Printed in China
10 9 8 7 6 5 4 3 2 1
04 03 02

*We dedicate this book to every barrel racer
who strives for clear communication
and a perfect partnership with a horse.*

Contents

Preface 9
Profile of Ed and Martha Wright 11
Introduction 17

  1. Breeding for Barrel Horses 18
  2. The Right Barrel Horse for You 28
  3. Conformation of a Barrel Horse 36
  4. The Mind of a Horse 50
  5. The Colt's First Year 58
  6. Starting the Prospect 64
  7. Beginning Barrel Training 80
  8. Increasing Your Speed 100
  9. Exercises to Improve Your Run 116
 10. Conditioning the Athlete 130
 11. Mental Preparation 136
 12. Hauling Your Horse 144
 13. Solutions for Common Problems 152
 14. Health Care and Nutrition 160
 15. Hoof Care and Lameness 174
 16. Tack and Equipment 180
 17. Great Barrel Horses We Have Known 194

Glossary 200
Association Addresses 203
Index 204

Preface

BY GLORY ANN KURTZ

Barrel racing is more than running a horse around the barrels. A successful barrel racer needs to know how to select a winning prospect and train him on the barrel pattern; how to ride a barrel horse that is already trained; how to take care of that horse to keep him healthy and sound; and how to keep herself or himself, as well as the horse, mentally and physically fit.

In "Barrel Racing: Training the Wright Way," Ed and Martha Wright will help you through all of these steps, just as they have thousands of students over the years. They have helped many beginning barrel racers become champions over the years as well as helped world champion barrel racers overcome problems with their horses or find a new horse. Their ability to match up a horse and a rider is uncanny.

The Wright's knowledge of both the horse's mind and body never ceases to amaze me. When talking with champion barrel racers from all levels of competition, the names of Ed and Martha often enter the conversation. Many times the Wrights have helped the barrel racer through a problem with their horse or they have owned or trained the horse the barrel racer is riding.

The telephone rings continually at the Wright household, as barrel racers — amateurs, NFR contenders and champions — are continually calling and asking for advice and suggestions. The couple have dedicated their lives to helping barrel racers and barrel horses. Their desire to help barrel racers, and their admiration, love and compassion for the horses are evident in each decision they make.

Profile

It's no accident that Ed and Martha Wright have trained more winning barrel horses than any other barrel racing team in history. From the time he was a child, Ed knew that he wanted to be a horseman. Raised in central Texas, he worked on ranches from the time he was seven years old, gathering goats, sheep and cattle on horseback. Ed had an uncle who roped and bull dogged; however, it was a neighbor, Jack Saunders, who was probably the most influencial person in Ed's life. Saunders owned a ranch close to Gatesville, Texas. He was a good man and a good horseman and he helped Ed a lot.

During high school, Ed competed in rodeos sponsored by the American Junior Rodeo Association (AJRA) and when he was 15, he won the Reserve Championship in calf roping and ribbon roping. Ed couldn't afford more than one horse at a time and never had a really good horse, but he had learned how to work with what he had. After high school, Ed attended Eastern New Mexico University at Portales on a full scholarship, majoring in physical education and agriculture. In 1971, he won the National Intercollegiate Rodeo Association (NIRA) Steer Wrestling Championship, as well as the steer wresting title at the NIRA finals. He also won the steer wrestling championship title twice in the Southwestern Region of the NIRA.

Martha's climb to stardom is well-known by barrel racers who have been around for a while. Martha was raised around horses and rodeos. Her grandfather, Everett E. Colborn was a well-known rodeo producer who produced rodeos at Madison Square Garden. Her father, Harry Tompkins is a rodeo legend — twice the PRCA All-Around Champion and a six-time Bull Riding Champion. Had Martha not taken an interest in living up to her Western heritage, she would certainly have been a disappointment to her father. But Martha not only took an interest in horses and barrel racing, she became a legend in her own right.

Most top barrel racers have become famous by virtue of one great horse that has carried them to many wins. In 1970 when 18-year-old Martha Tompkins showed up and won both go-rounds and the average in the Ladies Open Barrel Race at the TBRA Futurity on a four-year-old, black, double-bred King horse named Cowhand Breeze, many thought it was temporary fame, but Martha's success would not be short-lived.

Ed and Martha met in the spring of 1971 at a college rodeo. Martha was attending Tarleton State University and Ed was attending Eastern New Mexico University. That year, Ed was named National Steer Wrestling Champion and Martha was the National Barrel Champion at the National Intercollege Rodeo Finals riding "Breeze," the horse that many still consider to be one of the all-time great barrel horses.

A proud, 12-year-old Ed displays his prize after winning first in a horsemanship class.

In 1971, while still attending college, Martha got her Women's Professional Rodeo Association (WPRA) card and qualified for the National Finals Rodeo (NFR). Breeze had attended the NFR the year before, with Loretta Manual in the saddle, winning five go-rounds. During Martha's first NFR run, Breeze turned before he got to the second barrel, but that was his first and last mistake. After that, the pair won six of the 10 go-rounds and placed in nine. Martha was not only named the WPRA "Rookie Of The Year," but the pair set a new record for money won.

The year 1972 was a turning point in Martha's life. In January, she married Ed and only a few months later she lost Breeze to colic, at the tender age of six. If that wasn't enough, Martha had a full sister to Breeze that she had planned to take to the 1972 Texas Barrel Racing Association (TBRA) Futurity, but she too died, two weeks before the futurity.

Martha's barrel futurity career officially began in 1973 when riding a mare named Game Dame, owned by Jo and Tater Decker. She was Reserve Champion of the TBRA Futurity and the Houston Futurity (a futurity which only lasted two years.) In 1974 and 1975, Martha sat and watched the futurity as she was going to school full time getting a Bachelors of Science Degree in Biology.

Ed competing at a 1983 PRCA rodeo in Fort Worth. (Dudley Barker Photo)

Having been struck by the "futurity bug," Martha was back in competition in 1976, taking two mares to the TBRA Futurity: Poco Leo's Comedy and Marion Breeze, a half sister to Breeze. She placed high in the average on Comedy but Marion was no match for her legendary brother. In March of 1977, Martha went to work for Florence and Dale Youree at their ranch in Addington, Oklahoma. At the 1977 TBRA Futurity, Martha had the misfortune of hitting a barrel, but in the maturity, she placed sixth on Poco Leo's Comedy. She also ran the fastest time of the maturity on Youree's horse, Moon County, but hit a barrel.

In 1978, while still working for the Yourees, Martha took Grady County to the first Old Fort Days Futurity in Fort Smith, Arkansas, placing in the first consolation. Later that year, Martha and Ed went out on their own, training barrel horses. They took a mare named Miss Elegant Bar to the late futurities, as she wasn't quite ready for the TBRA Futurity. The pair ran the fastest time at the Oklahoma Cowgirls Association Futurity and finished fourth in the average. At the Colorado Futurity, they ran the second fastest time of the event and finished fourth in the average.

Now futurity contenders were starting to sit up and take notice. In 1979, Martha and Ed took two horses to the lucrative Old Fort Days Futurity — Fairbert, owned by Robyn Leatherman, and their own horse, Billy Bars Bug, nicknamed "Pokey." Fairbert ran the fastest qualifying time and finished third in the finals. Pokey hit a barrel in the trials. However, they took the same two horses to the Indiana Futurity, making a clean sweep! Headlines screamed about Billy Bars Bug winning both goes and the Championship, while Bert was named the Reserve Champion. Pokey also won both go-rounds and the average at the Colorado Barrel Futurity. That same year, he was named Reserve Champion in Junior Barrels at the AQHA World Show.

Billy Bars Bug was sired by Jerry's Bug, owned by Jerry Whittle of Dallas, Texas. Sired by Lady's Bug Moon, Jerry's Bug not only sired Billy Bars Bug, but started a streak of winning barrel horses for Martha, including Orange Bug and Top Bar Bug. During his Derby year in 1980, Billy Bars Bug won the Old Fort Days Maturity, and Martha won the consolation on Orange Bug. The winning kept on and during one 10-day period, from May 23-June 1, Martha made seven runs on Pokey, never placing lower than third, winning $7,122. Altogether, Martha won over $45,000 on the great gelding.

The year 1980 was a banner year for Martha, as she not only was winning on several Maturity horses, but she won the prestigious Texas Barrel Racing Association Futurity on Jetta Jay, a horse owned by Janice Saunders, that Martha and Ed had taken to train only three months prior to the Futurity, coming direct from the race track.

Through the 1980's, Martha and Ed's horses continued to stack up wins at major barrel futurities across the country. Ramblin Rally, was Reserve Champion of the 1986 Old Fort Days Barrel Futurity, then came back the next

A cowgirl at heart, Martha Tompkins took an interest in horses and barrel racing at a young age.

year and won the Derby. They ended up the year in the top 20 of the WPRA, besides winning and placing at several derbies across the country.

In 1989, Martha finished second on Major Movement in the Sweepstakes at the World Championship Barrel Futurity held in Oklahoma City, taking home close to $11,000. The following year they won the Sweepstakes at the big Lazy E Futurity, finished third at the high-paying Sweepstakes held during The Texas Barrel Race in Fort Worth and were Reserve Champions of the Heart of Oklahoma Sweepstakes.

Martha also barrel raced professionally on the gelding, winning the WPRA Texas Circuit and qualifying for the NFR. However, she chose not to take Major to the NFR as she was having trouble keeping him sound, and with 10 runs to go at Las Vegas, she decided instead to take her futurity horse to Oklahoma City. But altogether, besides a two-horse trailer won at the American Novice Horse Association Finals in 1992, Martha won over $100,000 on the great gelding.

Martha's 1989 futurity horse was Danish, a horse she describes as one of the "most fun" horses she's ever had. Danish came off the track with a 99 speed index. He won several futurities, including the North Texas Barrel Futurity, and placed at several others, including the Lazy E, San Antonio, Texas Arena News, Janet Meyers, Pineywoods and collected over $7,000 at the World Championship Barrel Futurity.

But the lasting memory of the year occured when they tipped over a $52,000 barrel in the finals at Fort Smith, catching the first barrel as they left it. They had run a 17.1, while the high-paying finals was won with a 17.3. The pair, however, went on to win over $32,000 in futurity and derby competition.

Even though there were several successful barrel horses between 1989 and 1992, Martha's next "great" horse was Jetta C Leo, sired by the great barrel horse sire, Jet Of Honor, and out of a Leo C Inman mare. The gelding won over $48,000 in barrel futurities, with his largest paycheck, $22,453, coming as Reserve Champion of the World Championship Barrel Futurity. He also won $12,701 for fourth at the Old Fort Days Futurity.

During his Derby year in 1993, Jetta C Leo won over $17,000, taking second at the Lazy E Derby and Elite Barrel Derby. He also picked up $7,521 for finishing fourth at the Old Fort Days Derby and $5,140 for fifth in the World Championship Barrel Derby. Altogether, Martha and the great gelding won over $100,000 before his untimely death in 1996.

Every year, Martha's name has been in the "leading riders" column of the *Quarter Horse News* Barrel Racing statistics, but 1987 was her best futurity year, finishing as the sixth highest money earner in the nation, with total earnings of over $52,244. While most of the riders above her on the list were strictly futurity riders, Martha continued to travel and win on the professional rodeo circuit.

During the 1980's, Ed and Martha purchased barrel futurity horses with their main criteria being that they had been raced on the track. They felt that since most barrel racers were 130-150 yards, the abilty to accelerate rapidly was vital to any winning barrel run. However, speed indexes were not the deciding factor, as the fastest barrel horse they had ever had, only had a 67 speed index. What was important was how fast a horse could run for a short distance — not where he finished the race.

However, in the 90's, Ed and Martha changed their criteria for futurity horses. They felt that the race track instilled too many bad habits in the colts and it took too long to train those habits out of them. So instead, they started to purchase younger colts that hadn't even been broke, and either broke them themselves or had them broke by someone they felt followed their criteria.

Then they started doing something unheard of in the barrel racing industry — breeding for barrel horses. They had been introduced to Jet Of Honor, a Jet Deck-bred stallion, that sired several top barrel horses, including the great Jetta C Leo, and started to purchase well-bred broodmares to cross on him.

Ed and Martha have become almost legendary in the industry by training, riding and winning on some of the nation's top barrel horses. Today they train on their 120-acre facility located just outside of Stephenville, Texas. When not on the road holding clinics from coast to coast, most of their days are spent in a lush, green pasture, with a plowed-up arena on one end, working their futurity colts and a few outside horses.

Introduction

BY ED WRIGHT

Martha and I make the ideal team. We balance each other. She finds all the positive traits in a horse, and I don't miss the negative ones. I start the colts, and get them familiar with going around the barrels. Martha finishes them, and does all the competing. Once I turn a horse over to her, we don't switch back and forth, unless the horse has a problem which she feels I can help with. Our keys to success are good judgment, education, and consistency.

Your horse's education begins when you go in the pen or stall to catch him. We want a horse to bond with us. We want him to be focused on us. It's just like raising children. You want their respect, not their fear. An educated horse can travel freely, stop, back up, pick up leads, and be soft and flexible in his body, with only subtle cues from the rider. Education for the rider means learning to use a sensitive, logical approach to handling horses. Seeing the horse's point of view, and helping him to understand and follow yours, is what good horsemanship is all about.

Successfully competing in any aspect of the horse industry requires a large investment of time and effort. We offer tools which will help you to use those resources more effectively. Our goal is to help you develop your horse to the point where he is automatic. Your horse may not reach the highest level of competition, but if you apply the training and conditioning techniques we've outlined in this book, you can be a winner.

"YOU NEED TO SELECT THE PEDIGREES THAT WORK MOST OF THE TIME."

1

Breeding for Barrel Horses

Charmayne James, 10-time World Champion Barrel Racer, bought her great horse, Scamper, for $1,100 from a feedlot worker after he had bucked off all the ranch hands. Sherry Cervi, the 1995 WPRA World Champion barrel racer, bought her horse from Shelly McLain, who found "Troubles" (Sir Double Delight) at the race track. He had been "tarped" and left on the ground when she first saw him, as he had bucked off all of his jockeys and had never made it to the starting gates. Shelly laid out $800 for the troublesome gray horse and later sold him for a suitcase full of money.

The 1996, 1997, and 1998 World Champion Barrel Racer, Kristie Peterson, paid $400 for her horse Bozo (French Flash Hawk), as an unbroke two-year-old, with intentions to just break the renegade and try to make a profit. Four straight NFR averages later, her early expectations definitely proved to be underestimations. Not only did Kristie win the World title and the NFR average, but on the way set a record time for 10 runs, as well as a record amount won for the year.

You can bet your bottom dollar that when these world champions chose their barrel horses, pedigree was not foremost on their mind. In fact, up until the last few years, most barrel racers didn't pay a lot of attention to the pedigrees of their barrel horses. It was financially unfeasible to own broodmares and pay a fee to have them bred to a stallion. Most barrel racers simply took a horse that was available and inexpensive and made a barrel horse out of it.

Later, when barrel racing started paying better, they became more scientific about the sport, realizing they needed both speed and set in their horses. They went to horse sales and bought race horses that didn't make it, cutting horses that were cheap, or just good, broke horses that they could train on the barrel pattern in the least amount of time.

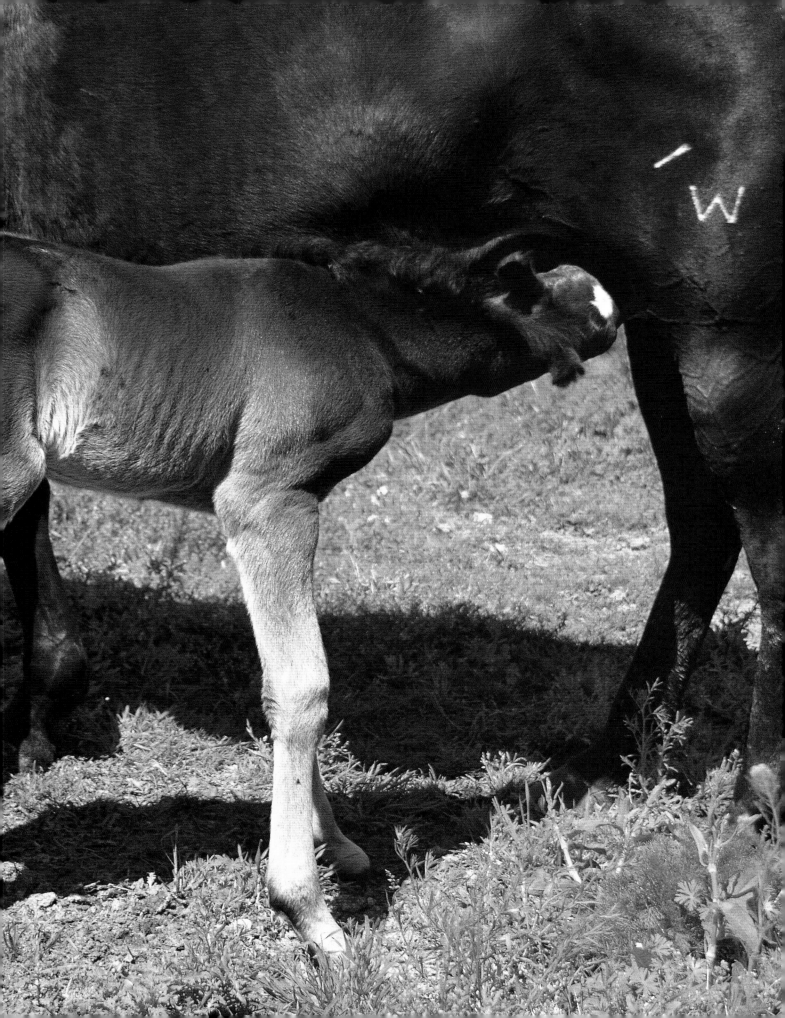

(overleaf) By raising their own colts, the Wrights are able to control the physical and mental makeup of their foals and produce more competitive barrel horses.

The growth of today's barrel racing industry, brought on by high-paying barrel races and futurities, has created a shortage of top-quality prospects. This has caused barrel racers to realize that having the proper bloodlines in barrel horses is just as important as having them in cutting horses, race horses, reining horses, halter or pleasure horses. They discovered that good pedigrees gave them the conformation, athletic ability and speed that was necessary in their horses to have them win in today's tough competition.

This is not to say that pedigree is the most important factor in purchasing a barrel horse. If we are considering purchasing a prospect from someone else, we watch a horse move, ride him and evaluate him on his own merits. Occasionally, we will overlook breeding if the horse has all the other qualities we look for in a barrel prospect. More often than not, the horse we think will make a great barrel horse has the pedigree to back up his ability.

When talking about the pedigrees of stallions and mares, we are referring to percentages, high and low. We are not talking about all of their offspring, just a high percentage of them, as there is no sire whose offspring all excel. Also, very few mares have all outstanding offspring. When studying pedigrees, you must not only be aware of the preferable "high-percentage of good offspring" pedigrees, but you also need to be aware of the "low-percentage" producers, so you can avoid them.

Foundation blood in a horse's pedigree such as from King P-234 (below) tends to add an even disposition and a heavier bone structure. (Photo courtesy of The Quarter Horse Journal)

There are pedigrees that work to produce a barrel horse a high percentage of the time, while there are others that don't. Therefore, anytime you're in the business, you need to select the pedigrees that work most of the time. In today's barrel racing industry, successful bloodlines usually go back to Thoroughbred breeding, just as they did years ago. However, it is common to see a dose of foundation blood on either the top or the bottom. This foundation blood, which usually goes back to such horses as Leo, Oklahoma Star, Wimpy or King, seems to add a more even disposition and a heavier bone so the horse's mind and body can stand up under constant racing and hauling.

Most of the early champions of the GRA (Girls Rodeo Association, which was the forerunner of the WPRA) rode horses that were foundation bred, with very little Thoroughbred blood in them. When we refer to a "cold-blooded" horse here, we are referring to one with foundation bloodlines, rather than race bloodlines. A "hot-blooded" horse would be one with a lot of Thoroughbred in it.

V's Sandy, the great horse owned and ridden by Jane Mayo, the 1959-61 World Champion, was sired by Oklahoma Star Jr by Oklahoma Star and out of Adonna by Bert by Tommy Clegg. That pedigree would fit any working ranch horse or cutting horse of the day. Sherry Combs Johnson, the 1962 World Champion, rode Star Plaudit, another horse sired by Oklahoma Star Jr, out of Sunshine Plaudit by Blondy Plaudit by Plaudit, another performance bloodline.

In 1970, Joyce Burk Loomis won the World Championship title on a memorable horse, Man O War Leo, a gelding sired by War Leo by Leo's Question by Leo. The gelding was out of Sooner Satin by Bojingles by Scooter Waggoner. Again, this is another foundation bloodline, carrying no Thoroughbred blood.

Gail Petska, who won the world title in 1972-73, rode Adobe Joe King, sired by King York by King Damon by King. The bottom of the 1964 gelding's pedigree was Tuno's Lady by Tuno out of an unregistered horse by Babe Grande. No hot blood there either. The next year, Jeanna Day rode Poco Excuse, a gelding sired by Excuse by Showdown by Wimpy, to the 1974 World Championship title. His dam's side went to Poco Soto by Poco Bueno by King. How much more foundation can you get?

Jimmie Gibbs Monroe's World Championship Barrel Racing title in 1975 ushered in the famed "Flit Bar" era. Jimmie's 1967 gelding, Robin Flit Bar, was sired by Flit Bar by Sugar Bars by Three Bars (TB). Flit Bar was out of an own daughter of Leo, representing the "magic" Sugar Bars/Leo cross, that has proven to be highly successful in the barrel racing industry. On the bottom side, Robin Flit Bar was out of Robin Hood Price by Osage Bob by Flying Bob, a race horse in his time.

Connie Combs Kirby, the 1976 World Champion, rode another foundation-bred barrel horse, a double-bred Joe Reed II stallion, Maudies Joak. Joe Reed II was sired by Joe Reed by Leo. Maudies Joak was sired by Joak by Joe Reed II and

was out of Maudie Leo by Leo. Carol Goostree's great Flicka Bob, the horse that took her to the 1979 World Championship, was race bred on the top, being sired by Bob's Folly by Three Bars (TB), but was foundation bred on the bottom, being out of Flicka Grande by Boco Grande IV by Burnt.

Martha Josey, who won the World Championship Barrel Racing title in 1980, rode Sonny Bit O'Both, a 1973 gelding that didn't have a bit of Thoroughbred or race blood in him. When the gelding was born, his dam was 22 and his sire was 16 and they were both foundation bred as far back as his pedigree goes on paper, tracing back to Peter McCue, Red Buck, Old Joe and Harmon Baker.

Even today, the most famous barrel racing horse of all time, Scamper, is probably the most "cold blooded" of all the successful barrel horse pedigrees. It is three generations before a well-known running horse appears on Charmayne James's great Gills Bay Boy's papers. The 1977 gelding is

sired by Gill's Sonny Boy by Sonny Gill by the race horse Tonto Bars Gill, which was sired by Three Bars (TB). On the bottom side, however, no running horse is evident in four generations.

History tells us that the Leos, Kings and the Flit Bars run barrels, work cows and rope calves. But to get the speed required today, it's helpful to cross those foundation bloodlines on those known for speed, like the Jet Deck-bred horses, which have proven to be the most successful bloodline in barrel horses of today. In fact, in statistics published by *Quarter Horse News*, Jet Deck has been the leading grandsire and maternal grandsire of money-earning barrel futurity horses for the past several years. Most of today's popular bloodlines trace back to Jet Deck, at least on one side of the pedigree.

Other Quarter Horse speed bloodlines that are popular today for barrel horses include Tiny Charger, Bugs Alive In 75, Johnny Boone and Go Man Go, if they are crossed on foundation-bred mares. Thoroughbred bloodlines that cross well with foundation-bred pedigrees include Fols Native, Beduino and Secretariat, which we predict will be the cross of the future. Secretariat's descendants seem to really put a lot of talent in a horse and they seem to have a mild manner to them. There aren't that many in the "cowboy world" yet and there aren't many sons or daughters in our area, but we think there's really some future with that bloodline.

The top barrel futurity horses over the past few years were sired by On The Money Red, Rare Jet, Merganser, Jet Of Honor, Jet OJ, Up N Truckle, Packin Sixes, Fols Blue Six, Easy Thistle, Easy Max, Shoot Yeah, Jet's Pay Day, Flaming Talent — all familiar names in the race horse industry, and with most of them going back to Easy Jet, Jet Deck or his sire, Top Deck (TB).

Some of the leading paternal grandsires were Easy Jet, Jet Deck, Bennie's Big Red, Dash For Cash, Fast Jet, Duck Dance (TB), Streakin Six, Six Fols, Top Moon, Flaming Jet, Special Effort, Bugs Alive In 75, Te N' Te, Raise Your Glass (TB) and Three Oh's, in that order. The futurity horses also are leaning toward race-bred dams.

However, if you buy a horse from certain racing bloodlines known for hot temperaments, or a horse with too much Thoroughbred blood, you had better be a patient, laid-back kind of person, because such a horse might have a little more "fire in the furnace," meaning he can be aggressive and hard to handle. Because of this, most novice riders should select a calm horse, which would not ordinarily have a lot of "hot" or running blood in it, but rather a foundation-bred pedigree, at least on one side.

On the whole, novice riders have been able to successfully train and win on Leo and Jet Deck-bred horses, so those pedigrees have proven to be outstanding. We already know that they have what it takes for a novice barrel racer to achieve his or her goals. However, before you purchase any young prospect, it's a good idea to do your homework and check on the prospect's siblings to see what they're generally like.

(opposite top) Jet Deck is the progenitor of one of the most successful bloodlines in barrel racing horses today. (Photo courtesy of The Quarter Horse Racing Journal)

(opposite bottom) Charmayne James bought Scamper, the most famous barrel horse of all time, for $1,100. No running horses appear on Scamper's papers for three generations on the top side and four generations on the bottom. (Kenneth Springer Photo)

The sire is a very important element of your horse and we feel the sire gives his offspring heart, determination, attitude and conformation in that order. However, horsemen from every discipline agree that the dam possibly determines from 80 to 90 percent of the attributes of the resulting foal. We personally feel that the dam dictates a lot of the conformation, attitude, ability, heart and desire of the foal, in that order. Also, colts spend a lot of time with their mothers and they pick up a lot of her attitude and traits. The mare trains the baby. If the mare is quiet and easy going, the colt will have more of a chance to be that way.

No one can go by what a stallion or mare has accomplished, what they look like and what their breeding is and totally know what they will produce. So you need to check on the offspring of both the sire and the dam before you either purchase a colt or breed for one. However, occasionally you will find a stallion that is a "strong breeder," which means he usually puts more of his traits and conformation in the offspring than the mare does.

One stallion in the barrel industry that we feel did this is the now-deceased stallion Jet Of Honor, sired by Jet Deck out of a daughter of Lightning Bar by Three Bars (TB), which was out of a daughter of Leo. Lightning Bar was a race horse but he sired Doc Bar, the great sire that created today's cutting horse dynasty. Jet Of Honor was truly a prepotent stallion and we have to give him equal billing with the broodmare, because he definitely put speed in the foal, as well as heart, try, and the ability to run barrels, regardless of what he was bred to. Jet Of Honor sired a better-looking horse than he was and that's the sign of an outstanding sire.

We've never known of a sire that has sired colts that can work so many different styles for so many different levels of riders and trainers. Because of the attitude and trainability of the Jet Of Honor offspring, they make top barrel horses. Some of his great offspring include Jetta C Leo, the great gelding that Martha took to over $100,000 in earnings; Rene Dan Jet, AQHA World Champion Barrel and Pole Bending horse; Band Of Honor; Mr Honor Bound; Rudy Val; Sexy Classy Moon, and Easy Honor Jet, to name a few.

We can study pedigrees and conformation, but until we actually see a colt sired by a certain stallion and out of a certain mare, we never know what they will produce. Also, you must be aware that even though a certain combination has produced a winner the first time, it doesn't mean it will happen again. You may have a better chance the second time, but it's never a "sure deal," and you have no guarantee.

That's why breeding is a very tough subject to learn. No matter what research we do, there are still many variables in breeding. We feel, among other things, that the age of the broodmare is important. If you breed a mare that is from 3-6 years old, she probably won't produce nearly as well as she will if she is 7-10 years old. Then, after 10 years of age, the mare's productivity goes down a certain degree each year, depending on her health and the care given to her throughout her life. Because of these variables, determining the best breeding strictly on paper (by the horse's pedigree) isn't highly accurate.

As tough as barrel racing has become, it is imperative that you have as few negatives as possible in a barrel prospect. There are more barrel racers today than ever before, and they're breeding and buying better horses. That's why barrel racing has become tougher than it's ever been. Consequently, we are starting to raise some of our own horses. This way we can control the quality of our foals, both their physical and mental makeup. Also, it lets us know what background and education the horse has. We will still buy some outside colts, but we want to control our destiny as much as possible. We feel that if we raise a colt, we know what has gone on with the horse, physically and mentally, throughout its life.

To produce a barrel racing prospect, the broodmare should not only be outstanding on paper, with a top pedigree, but she should have good conformation, physical ability, a good attitude and intelligence. If she has speed, it should be short, quick speed — from 50 to 100 yards. Speed is definitely a factor in barrel racing, but we've seen some very competitive barrel horses which were just average to above average on speed. Try and heart mean a whole lot. For that reason, we have broodmares that go back to Leo, King P-234, Jet Deck and some old foundation horses, race horses and cow horses.

A broodmare can be a great individual. However, she should be chosen by what she can produce rather than what her own abilities or speed are. Therefore, when we pick a broodmare, we look at more than just her pedigree, conformation, attitude and looks. We want to see a colt out of her.

When we own mares that we have ridden colts out of, and have them bred to the same studs, we have a better chance, percentage-wise, of getting another outstanding colt. But we still can't control that 100 percent of the time. If a mare has already produced a barrel horse, or one that could be a barrel horse, chances are she could produce another one. On the other hand, if you don't like her other colts, we wouldn't suggest you take the gamble on breeding her for another one.

When we pick a broodmare, we want her to have some substance and bone. In other words, if she has small bones, she probably won't be stout enough or have the strength she needs, and will be more susceptible to unsoundness. She could pass this on to her offspring. We want her to have good-sized, strong bones, but not an over abundance of bone, like the large bones of a draft horse.

Since we want the mare to produce colts that have good conformation, she needs to possess the same conformation as the ideal barrel horse. Some broodmares produce colts that have better conformation than they do, which is the sign of a great mare. So if she doesn't have the conformation you desire, but you've seen her colts, and she outproduces herself, that's okay. Remember, there are

(above and right) Because colts spend a lot of time with their mothers, they will usually pick up a lot of her traits. Easy-going mares tend to have easy-going babies.

always some good, or even great horses that you would never expect to excel when you look at their bloodlines.

Don't select a broodmare or a stallion that has a major flaw. We do not want to get in a position of "making up for weaknesses" in either the broodmare or the stallion. We want to strive for perfection as much as possible in the mare and the stallion. However, sometimes things must be put on the scales, and you may have some good attributes in a horse that outweigh some bad ones.

As an example, Jet Of Honor had bad feet. He has sired colts that have had good and bad feet. But still, one should never breed to a horse thinking he or she can overcome one of his weaknesses, or use a broodmare, thinking the stallion can overcome one of her weaknesses.

If a stallion or a mare has a weakness, more often than not, that weakness will come out in the colt no matter what you do. Therefore, you need to put attributes on the scales. If you were breeding to Jet Of Honor, speed, a good mind, perfect conformation and athletic ability had to outweigh the colt having good feet. However, we wouldn't breed a mare with bad feet to a stallion with bad feet, because then your odds of the colt having bad feet have just gone up drastically.

Breeding for barrel horses today is financially feasible due to the large purses and the fact that outstanding, well-bred prospects are hard to find, and very expensive if you do find them. Also, by breeding, you have better control of your destiny; that is, if you're educated about what it takes to produce a good barrel horse.

Breeding for barrel horses is not for everyone. You must have the patience and financial capability to wait. First you need to buy a broodmare, have her bred, and wait for a year to have the colt. Then you have to train it, starting when the colt is born with imprint training and going through all the other stages along the way, until it is ready for you to compete on.

You must also have an appropriate facility to run broodmares and raise colts. If you plan on owning a stallion, that will take additional time and different facilities, as well as the expertise to be able to handle a stallion, or the financial ability to hire someone who does. So think twice before getting into the breeding end of the business. It's definitely a big, and expensive, commitment.

"YOU MUST REMEMBER THAT EACH HORSE IS AN INDIVIDUAL."

2

The Right Barrel Horse for You

When selecting a barrel horse, you must be open and honest with yourself. Ask yourself: "What caliber of horse am I capable of riding?" "How much can I afford to spend?" "What are my goals after I've purchased the horse?"

First, pick a horse that fits your capabilities. Many people don't realize their limitations. They want a horse which has talents they can't match. You need to know the type of horse you get along with. For example, have you had more luck with aggressive horses or laid-back ones? Your choice should not be determined by what you think you like, but what you're capable of handling.

If you're a novice rider looking for a first-time horse, we suggest you look hard at an older, solid barrel horse that can help train you. If you are a successful barrel racer looking for a futurity prospect, it's imperative that you choose your prospect carefully. It takes two years to train a barrel horse, and there's a lot at stake because barrel futurities pay out hundreds of thousands of dollars in prize money.

When we look at a horse, we first try to determine if he has athletic ability. We want to see him standing still, trotting toward us and away from us and then turned loose in a round pen. When the horse is trotting toward us and away from us, we are usually looking for the movement of his legs, making sure he travels correctly. We want the feet to have a lot of action in them, forward and back, but no action straight up in the air. We don't like elevation of the hock, knee or hoof. We like a horse to keep his feet, knees and hocks low when traveling. They should clear the ground, but be as close to the ground as they can be. Excessively high movement is wasted motion in a barrel horse and takes away from speed.

When Ed tries out trained barrel racing prospects, he wants a horse which handles turns by getting on all four feet and moving forward at all times with a little flex to his body.

You can tell if a horse is a good mover by having an assistant trot him away from and toward you. His legs should travel correctly and his feet should have a lot of back and forth action with no up and down motion.

When the horse is turned loose in a pen, we'll step to his hip or a little behind him. That's how you get a horse to go forward. As he moves around the pen, we want to see his top line move smooth and level, and his head carriage to be soft and supple, not jerking up and down. The whole body should move as a unit — that means each quarter, both shoulders and both hips, working individually, yet moving in time with each other.

We don't want a horse to hit the ground hard. He should travel with a smooth, even motion, like he's a soft, supple athlete. We like to see long, fluid strides, which tells us he can more than likely really move out. We like to see a horse extend his back foot print over his front foot print, which tells us he has a long stride.

In the round pen, when we step to the horse's shoulder, we can slow the horse down with the position of our body. If we step in front of the horse, we stop him. When we stop him or slow him down, he should be able to shorten his stride. This tells us that in a barrel turn, he can hold the turn better and leave it quicker. A long stride in a barrel turn does not control body position like a short stride does. A short stride around the barrels makes for a quicker turn, but he needs the long stride to run between the barrels.

When he turns around, we want him to spread his back feet and pull with his front feet, using all four feet against the ground. We don't want to see a horse that turns around by swinging his front end in the direction of the turn and keeping his back feet close together. He can execute a turn in a round pen that way, but around a barrel, he needs to spread his feet to hold the ground.

While working with the horse in the round pen, we are not only judging his physical ability but also his mental attitude. We like it if he listens to us when we are trying to move him forward (which is the easiest part), stepping to his shoulder to slow him down and stepping to his eye to stop him. If he stays thoughtful and attentive of where we are, by the movement of his eyes and ears, we like that better than if he just gets wild-eyed and tense and wants to run over us or the wall.

To a certain degree, you can also read the "try" and "heart" of the horse in the round pen. If he stays soft in his mind and just keeps moving, even if he's tired, that shows some heart and try. If the horse gets nervous when he starts to get tired, that shows that he may not have much heart, but for sure, he doesn't have the mind that we want and need for a barrel horse.

A barrel prospect should have three traits: conformation, breeding and mind. We can live without conformation and breeding to a certain degree, but we can't live without the mind, which includes heart, try and desire. These are also the hardest to read. After looking at the horse's conformation and movement, we get on him in a round pen if he is started under saddle. The round pen allows us to have more control of where he goes. If he is far enough along, we will go to a larger pen or an arena, and eventually to a pasture.

We'll trot and lope him in circles, going both directions. When we're riding a horse, we want the shoulders, hips, head and neck to all be in balance. A horse can be fairly athletic, but it can feel like his front end is doing one thing and his back end is doing something else. The motion should fit together. Just like a horse should look balanced, his movement should be balanced.

A cheetah makes a motion just like we want a horse to make. They're supple and soft, and everything looks fluid when they move. "Fluid" means motion that has timing. There's timing between every part of the body that's moving. We don't like a horse that jars you or himself when he hits the ground.

We also don't like a horse that moves with his back feet inside the front feet when he's traveling. That type of horse can't run, stop or turn around correctly or quickly. That's the way a dog travels — when he trots toward you, his back feet will travel inside his front feet. A horse that is narrow behind and wide in front may be able to travel in a straight line without interfering, but if he is making a run, he will usually interfere with himself by overreaching (reaching with the back foot and clipping the front foot or leg). Also, we never want to see the horse's head jerk up and down when he is traveling. When he bobs his head, it usually means soreness, but it can also mean that he is a poor athlete.

To check a horse's mind, we like to start out with the horse totally cold, taking him right from the stall or the pen. We warm him up, then ride him until he gets slightly hot, just to see how his mind works. If the horse is too young to be ridden, we lope or trot him in the round pen until he just breaks a sweat on the neck, behind the ears or between the hind legs. This stresses him a little, and when the stress hits him, you can read the other side of his mind — the side that wasn't evident when you first put him into the pen when he was fresh and cool.

A horse without a good mind won't slow down very easily, even after you have educated him to your body motions and what they mean. He'll still want to run over you. If he get's flighty or wild-eyed when you try to stop him, he doesn't have the kind of mind that you should be looking for in a barrel horse.

To warm a horse up and put a small amount of stress on him, ride him at a long trot and lope, then slow down, turn him around and handle him until you can figure out if his mind stays soft under a little bit of excessive work. Here again, ride him until the sweat appears on his neck, in back of his ears or between his hind legs. How a horse reacts when he is tired and stressed is the true test. However, don't ride him until he is frothing with sweat. You have to use common sense, because you can make any horse crazy if you push it beyond its limit.

We don't like a horse that is lazy, but we also don't like one with too much "go." Some lazy horses, however, make top barrel horses. We had one little mare that was so lazy that you had to carry a stick to get her to trot in the pasture. Because of that, we sold her, but she turned out to be one of the best barrel horses around. She's really lazy when you exercise her, but when you go down that alley, she knows where she is and she's all business.

We like a horse that senses what we want when we ask for it. If we relax and ask him to drop his head, we want him to travel quietly. If we move forward in the saddle, we want the horse to come alive. We sure don't want a crazy, wound-up, chargey horse; but on the other hand, we don't really like a horse that you've got to hustle all the time either.

When you buy any horse, you need to determine if he is merely broke or if he is educated. To us, a broke horse is one which will let you get on and plow rein him around. He is usually one trained by intimidation. We prefer an educated horse which, if he is taught correctly, will pick up leads, stop, and move his shoulders by responding to leg and rein cues. He also flexes at the poll and rounds his back in true collection. His rider has no trouble keeping him balanced on all four legs.

If a horse is educated correctly, you can position any part of his body from head to tail, which is what you need when starting your barrel pattern. Breaking, or intimidation training, sometimes creates problems that will be hard to overcome. Usually these problems don't surface when you are working the horse slowly; however, when you start going fast, they quickly surface.

We want to know if the horse has a problem that can't be overcome, because we work in a speed event. With a speed event, even with an educated, soft horse, there's a certain amount of stress. With stress, problems that have been educated into the horse can surface much more easily than with quiet, easygoing riding. Also, if a horse is educated totally different from what we want, that's the same as wrong, as far as we're concerned.

If a horse has been started on barrels, we like to lope him around the barrels, just to see how he turns. If he is really educated, we can tell, to a certain degree, how he will take a barrel within the first 30 minutes that we ride him. We may not, however, be able to tell how fast he is, if he will be able to beat the clock, or what he's going to do competitively later.

If he goes to the barrel and has a positive attitude about it, rating or getting into position (even if the rate's in the wrong place or the wrong way to fit my style), he shows me that he possesses desire and ability. All of these must go together. If he wants to drop his front end and start his turn too early, or he wants to rate way too early, you need to pick him up and handle him with your hands, legs and body, so that you can tell if he can be changed or if he's so set in his ways that you're going to have to do some extensive educational changes.

When you get him warmed up and you ask him to do these changes, and he does them, then he's probably susceptible to change. However, remember that the hotter he gets, the more he's likely to go back to his old habits. That is where you must make your judgment call. This is tough — just like reading the horse's mind, try and physical ability, although physical ability is easier to read than try and desire and the horse's mind.

If a horse runs to a barrel and you feel his mind suddenly become scared and hard, you feel his body get tense and electric, and you can't control him, that is bad. You must decide right then if you can change that horse and if he will stay changed, even when someone else rides him. Our goal is to not only change the horse so that we can ride him, but to educate him so that someone else can ride him. We would have to pass this horse up because he is not a horse that we would be proud to sell and that would advertise for us.

When trying out a trained or partially trained barrel horse, you must remember that each horse is an individual. We let some horses turn flat, while there are some horses we let really get on their hindquarters. A horse that turns flat doesn't bend in the rib cage, backbone, neck or head. They remain extremely straight through the vertebrae. The vertebrae doesn't have any flex. A lot of flat-turning horses also have a tendency to drop their front end a little bit in the turn.

A horse that gets on his hindquarters is one that, when he runs to the barrel and in the turn, drops his rear end lower than his front end. His front end can be elevated to the degree that he is even light in the front feet — the front feet are not in the air, but light. In other words, the horse is on his rear end and then pushes out of the barrel on his hind end.

The ideal, athletic horse, however, gets on all four feet and handles the turn, mobile forward (moving forward at all times) with a little flex to his body. This is a horse that is as close to the ground as he can be with his belly, using all four legs, pushing with his rear end and pulling with his front end in the turn. He uses each leg separately and what's above it — the front legs, shoulders, back legs and hips — and divides all the motion among the four legs, 25 percent to each.

The horse must have a certain amount of rate or slow-down, which is necessary to gather himself up to hold the turn perfectly. You'll lose some of the turn if it's too slow; you'll lose to a faster horse. We want a horse that doesn't have a pause behind the barrel. That's why a horse must use all four feet. In this motion, the rear end is used a little more in the first few jumps away from the barrel, but we still want him to use all four cylinders (four legs, two front shoulders and two hips) as equal as possible in the turn.

A horse overuses his rear-end when he does not use all of his physical capabilities, enabling him to excel in a turn. If he drops on his rear end, or drops on his front end, there is a point in the back of the barrel where there is a pause, and then he has to start running again. Running level, with all four feet on the ground, is much easier for the horse and he can stay mobile, (moving forward), which helps you win.

Proper conformation is another important factor when selecting a barrel horse. Remember, however, there are exceptions to every rule. We may have a horse that looks like a "billy goat," but makes a great barrel racer. Although, 99 out of 100 such horses will probably be nothing. We know there are freak situations in breeding, conformation and motion, but they are very rare. We discuss conformation and movement in depth in Chapter 3.

The price of a horse is also an important factor when selecting a prospect. Remember that the "least expensive horse you buy is the high-quality horse. And the highest-priced horse you buy is the one that has no quality." In other words, even if you can pay lots of money for a good horse, in the long run that's really not an expensive horse because you're getting a lot for your money. You're getting a good return on your investment. On the other hand, when a horse has no quality, it doesn't matter what you pay for him, it's too much. You didn't get anything for your money. That makes him outrageously expensive. Only one out of 50 low-quality horses will ever make it as a barrel horse. With a high-quality horse, we find that 48 out of the 50 will make some level of barrel horse.

Once you've tried a horse out and like him, before you spend your money for either a solid barrel horse or a prospect, be sure to have your veterinarian check the horse out. You may be able to see conformation flaws, but there are a lot of things you can't see that can be wrong with the horse.

If a horse overuses his rear end or front end when turning a barrel, he will lose time at the back of the barrel where he has to pause before he can start running again.

This could mean a bone spur in a joint, such as a hock, ankle, knee or stifle. With good radiographs, your vet can check for spurs. Your veterinarian should know how to feel shoulders, backs, or hips to find atrophy or muscle deterioration. The horse should be backed up to within six feet of a fence, with his hind end facing the fence. The vet should get up on the fence and look down on the horse's back, hips and shoulders, and make sure there is nothing that is out of balance on both sides. A hip could be shaped a little bit different on one side or the other, or one hip could be higher than the other.

Your veterinarian should also look inside the horse's mouth for any kind of deformity in the bone or with the teeth. He should look for cuts on the tongue or if there are any teeth knocked out. The bars of the mouth should be checked for scar tissue. Always have a veterinarian check the horse's eyes, not only for white spots, which would be the result of a blow, but also for cataracts or any deformity in the eye.

An equine specialist looks for motion. The way the horse travels can indicate soreness in some area. The vet should have an educated eye for motion which can help him go to the area that he thinks needs to be checked. Always have your vet draw blood on a horse before you purchase it. If the horse has an elevated muscle enzyme count, he could have a tendency to have his muscles "tie up." If he has an elevated white cell count, he could have an infection somewhere. Tests can also be made to determine liver or thyroid malfunctions.

A vet should also check to see if the horse has a breathing problem or a heart problem. The heart should be monitored standing and after activity. Both are very important. A blood count also has to be taken at two different times — while the horse is standing quietly and after he is in motion. Besides giving your horse a complete physical, have the vet x-ray the horse's legs and feet. You've heard the saying "no foot, no horse." That is very true, but with barrel horses it's "no legs, no horse."

"A HORSE'S STRIDE CAN MAKE OR BREAK YOU."

3

Conformation of a Barrel Horse

When Kristie Peterson, the 1996, 1997, and 1998 World Champion Barrel Racer, found "Bozo" in a pasture and purchased him as a gangly, broncy two-year-old colt for $400, pedigree and conformation were not on top of her list. Her thoughts were on whether or not she could get him broke and sell him for a profit. Obviously, she got him broke, and in 1996 alone, she earned over $170,000 on him. After that, he was probably not for sale at any price.

There have been lots of world champion barrel horses that were purchased for little or nothing, but whether the barrel racer knew it or not at the time, most of them turned out to be balanced horses with trainable minds. If their conformation was not balanced, their legs and bodies couldn't have taken the pounding they had to take when they were hauled thousands of miles a year. If their minds hadn't been responsive, they wouldn't have been willing to run barrels day after day. Of course, there's always an exception to every rule. However, these are two attributes that most top barrel horses have in common. And while both conformation and a good mind are equally important, we will discuss conformation, along with other physical attributes of the barrel horse, in this chapter.

Proper conformation, in most cases, determines the athletic ability of a horse. If we had to pick one certain aspect of what we want in a barrel horse, we would pick a horse whose body was balanced with correct shoulder and hip angles. Good conformation is more important in a barrel racing prospect than it is in an older, seasoned barrel horse. If the solid barrel horse does not have the ideal conformation, it is "after the fact," as that horse has already proven he can be a top barrel horse, even if he does not have ideal conformation.

However, with the cost of training a barrel horse, it is better to start off with a prospect that has the conformation most likely to make him a good athlete. Also, even though good looks are certainly not essential to a top barrel horse, it is always helpful to have a good-looking horse when you decide to sell it.

HEAD

The way a barrel horse's head looks is not that important. We feel the most important parts of the head are the horse's eyes and mouth. A horse's eyes should be fairly good-sized and have an intelligent look. We feel that a horse with small eyes, often called "pig eyes," doesn't have as good vision as a horse with big eyes. We don't have any documentation for this. It is strictly personal opinion that is based on years of experience. We feel that a horse with big eyes can see more and is less afraid of what he can't see. Also, a lot of width between the eyes is usually the sign of an intelligent horse.

We like a horse to have large nostrils since small ones can't take in as much oxygen. The more oxygen a horse can get easily, the better. We feel that most horses with large nostrils seem to have more try about them. Again, this is just our opinion, but it is based on many years of observation of many different horses.

A short mouth is more advantageous than a deep mouth. When measuring the depth of a horse's mouth, measure on the side of the horse's head from the muzzle to the end of the mouth opening, which is located up the jaw line. A horse with a deep mouth seems to have more bit problems. The bit could interfere with his teeth. Instead of sitting on the bars (interdental spaces) between the front (incisors) and the back teeth (molars), where bits are supposed to lie, the bit could bang against the horse's teeth, causing irritation and pain. Also, deep-mouthed horses tend to be harder to educate, as they seem to have less sensitivity or feeling in the mouth than a short-mouthed horse.

However, a really short-mouthed horse is not good either. The corners of the mouth should be where the bit lies in the mouth, behind the front teeth and in front of the back teeth. Extremely short mouths are deformed by shape and are very rare, usually with a "phony" feeling. (A "phony" feeling means a horse will be hard to educate because of extra sensitivity.) Of course, we don't like horses with deformed mouths, such as one with an overbite, with the top teeth coming over the bottom ones (parrot mouth), or underbite, where the bottom teeth come over the top teeth. Both of these can cause eating and chewing problems. The top and bottom teeth should meet evenly, with neither lapping over the top of the other.

The ears should have a good, wide set to them. We don't like a horse that is pin-eared, which is when the ears come out of the head unusually close together. From our observation, horses with that trait are usually nervous and have a short attention span. Here again, we're talking

(previous pages) Barrel horses need good conformation and balance to withstand the pounding they must endure over the thousands of miles they are hauled each year.

(opposite) A good head on a horse means good-sized eyes with width between them, wide-set ears, large nostrils, and a short mouth.

A horse with big eyes not only has better vision than a horse with small eyes, but also has a more intelligent look.

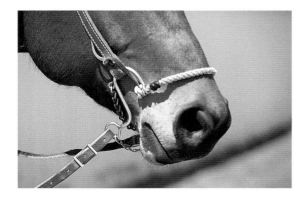

A horse with a short mouth has fewer bit problems and is generally easier to educate than a horse with a deep mouth.

percentages. Not all pin-eared horses are nervous, but we have seen it happen a lot. We're not talking about ears that are just close together, however, but rather a deformity. Any deformity scares us because for some unknown reason, it usually adversely affects the mind.

NECK

The neck of the horse should have some length to it but it should be balanced with the rest of the horse's body, not extra long and not extra short. If the horse has a shorter back, he can have a little shorter neck. If he has a longer back, his neck should match and be a little longer. It has to balance.

The neck should tie in cleanly at the throat latch, which shouldn't be thick. It should be clean (tight and small, not full and massive) behind the poll, which is where the neck ties into the head immediately behind the horse's ears. It should also be refined, where it comes into the chest. We definitely don't like a crested neck (a rise of fat or a thickening of the neck where the mane is) and we don't like a horse whose neck ties in really thick in front of the withers. The ideal throat latch is small and refined, even on a stallion that has a large, muscular jaw.

A good neck balances well with the rest of the horse's body. A horse with a short back should have a shorter neck. The neck should tie in cleanly at the throat latch.

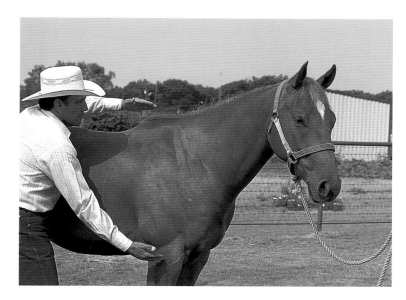

A long shoulder and an average heart girth are preferable for a barrel prospect.

SHOULDER

We like a long shoulder, with the point of the shoulder as low and forward as possible. The hip and shoulder angles should be symmetrical. When the angles match, you have a more athletic horse, one that is capable of being a barrel horse. Matching angles on a horse allow the shoulder and hip to move compatibly with each other, which makes a better athlete. If the angles of his shoulders and hips match, the stride will also match, which means that all parts of the horse will move as one.

A shoulder should have some muscle, but we don't want a horse that is too wide between the front legs and over muscled in the front end. If a horse is massive and wide in the front end, he usually does not move balanced on all four legs. Instead, he relies on his front end excessively, which causes him to lose speed around the barrel because he has to exert more effort to move his over-muscled front end around the barrel. A refined front end does not require as much effort, allowing the horse to move in a soft, athletic motion around the barrel.

FRONT END

We like to see the shape of an inverted "v" between the front legs, rather than a boxy, square or flat front end. The "v" is the musculature between the front legs, where the legs attach to the horse's chest and shoulders. This "v" makes the horse a supple, more balanced horse and lighter in the front end. The wide, boxy front end makes a horse hammer the ground, which is hard on his bones and muscles.

An athletic horse should have a lot of muscle on the inside of the front legs that taper from the chest all the way down to the knee. We want the horse to have a bulging muscle, but we want that muscle to be extremely long also. Short bulging muscle is not athletic muscling. It doesn't move as well and is much more susceptible to soreness. Longer muscling allows a horse to move better and makes him less prone to soreness and injury.

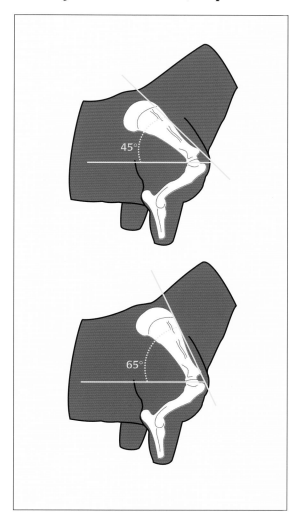

(figure 3.1) Long, sloping shoulders allow a horse to stride farther than do short, steep shoulders.

A refined front end allows a horse to move in a soft, athletic motion around a barrel. The Wrights like to see an inverted "v" between the front legs.

The knees should be close to the ground, which means the cannon bones should be as short as possible. An athletic horse, with short cannon bones, moves in a low, more athletic, manner. When a horse exerts himself, or is running the barrel pattern, the leverage created by long cannon bones puts stress on the neck, shoulders, back, hips and forearms, which affects the soundness of the horse. Short cannon bones help the horse's athletic ability, since the leverage created by running spreads the stress over the entire body, keeping him sound longer.

Also, a cannon bone should have good width to it because sufficient width in a cannon bone usually denotes a bone with good density. A small cannon bone, a high percentage of the time, is weak. Since we can't actually check density of a bone, the larger cannon bone, looking from the front or the side, points to the fact that the bone is more dense. It definitely will be more durable and stand up longer for the barrel horse. When viewing the cannon bone from the side, it should also look wide, with the bone, ligaments and tendons stacked in such a way that the leg has a wide, flat appearance. There should be room behind the cannon bone for the tendons and ligaments.

It is important to have a horse's shoulders, knees and feet line up. When he runs, pulls or pushes away from a barrel, there should be nothing out of line to pull one of these body parts improperly to the side. If something is out of alignment, it puts stress on the inside or outside of every joint, from the shoulder joint to the ground. Obviously, you could have a soundness problem if these body parts do not line up. But at the least, a horse will be exerting harder to accomplish the job.

You can check this out yourself by holding a string with a weight tied to the end of it at the point of the shoulder. That straight line should divide the leg equally when viewed from the front. If the string runs down the center of the horse's shoulder, knee, fetlock and foot, you have a horse that lines up well. Then, from the side, drop that string down from the middle of the shoulder. It should equally divide the front leg down to the fetlock. The bottom of that line should end directly behind the heel of the foot.

It is important that the horse's knees are not behind the string. That's referred to as being calf-kneed or "back in the knees." Horses with this type of leg conformation can have a lot of joint problems. Calf-kneed horses are not usually sound very long. On the other hand, a horse shouldn't be over in the knees either. In this case, the knees would come out in front of the string. But, if we had to choose between a calf-kneed horse and a horse that is over in the knees, we would take the latter. When a horse is back in the knees, moving causes a lot of concussion and trauma on the front joints of the knees, because the joints do not sit properly. This problem increases the chance of injury. The buck-knee or over-in-the-knees horse has a lower chance of injury.

We don't like a horse that toes in or is pigeon-toed. Toe-in means the toes point toward one another when viewed from the front. It can be congenital (acquired in the uterus and not inherited), and the limb may be crooked from the chest down or from the fetlock down. When a toe-in horse moves, he has a tendency to paddle with his front feet. His legs can interfere with each other, especially at the fetlock joint, causing damage to the other leg.

By the same token, we don't like a toe-out or splay-footed horse. That condition is also usually congenital and is usually due to legs that are crooked from their origin (at the chest) down. A horse that toes out will usually wing to the inside when he moves. Both toe-in and toe-out conformation may be controlled or partially corrected by corrective trimming and shoeing, but only in the very young horse. Attempting to correct these problems in older horses, whose bones are set, can cause more unsoundness than the defect.

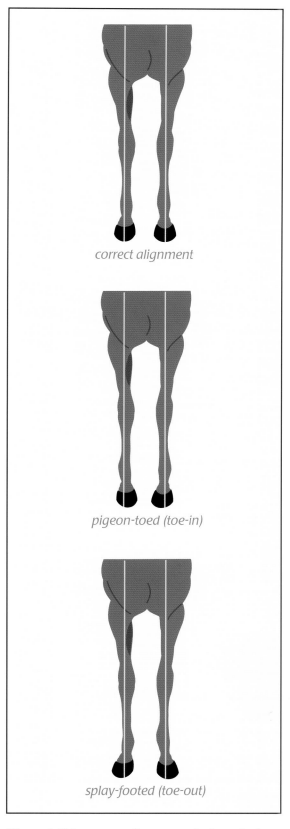

(figure 3.2) Improper alignment of a horse's front end puts undue stress on the joints and can lead to soundness problems.

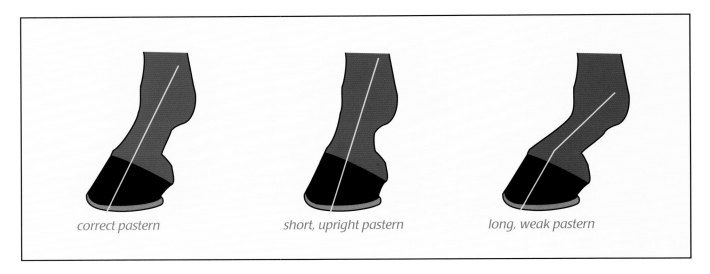

(figure 3.3) Barrel horses require correct pastern conformation. Short, upright pasterns and long, weak pasterns can both lead to soundness problems.

PASTERN

The pastern should be of average length. A very short pastern takes a lot of concussion from the ground, which is hard on a horse, and hard on his ability to stay sound. But a long pastern is worse, because it has too much give to it when the horse steps and that stresses the suspensory system with every step. The suspensory system includes the tendons and ligaments in the legs.

HEART GIRTH

A horse should have an average heart girth. A horse with a really deep heart girth uses his front end too much and, as we have previously mentioned, we want a horse with four-wheel drive for barrel racing. An exaggerated, massive heart girth causes a horse to not be as mobile as he needs to be to run barrels. On the other hand, we don't want a "lizard-gutted" horse (one that is shallow in the heart girth or tucked up like a greyhound). Such a horse doesn't have the power to push when his body is bent in the middle of the turn around the barrel, and this makes him vulnerable to soreness and injury.

WITHERS

Good withers hold a saddle in place. A horse with a round back or mutton-withers is more difficult to fit a saddle to and it's hard to keep the saddle from rolling around. Also, we've found that horses with rounded withers usually tend to be heavy in the front end. But don't judge a prospect's withers too early. Sometimes a young horse tends to be round and pudgy in the withers, but when he grows into a three or four-year-old, he develops a good set of withers.

BACK AND UNDERLINE

A horse should have a short, strong back. A long back is weak. However, to compensate for a long back, a horse has to be strong over the loin. A horse also should have a long underline, which is the distance along the bottom of the belly, between the front legs and the back legs. A short

CONFORMATION OF A BARREL HORSE

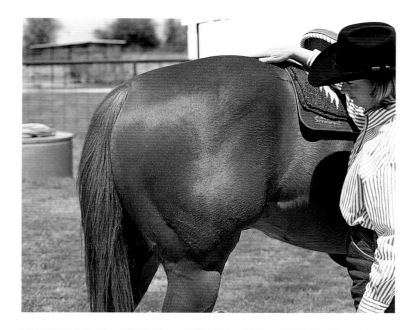

Hindquarters should be long and neither flat on top nor slope off too much at the tail.

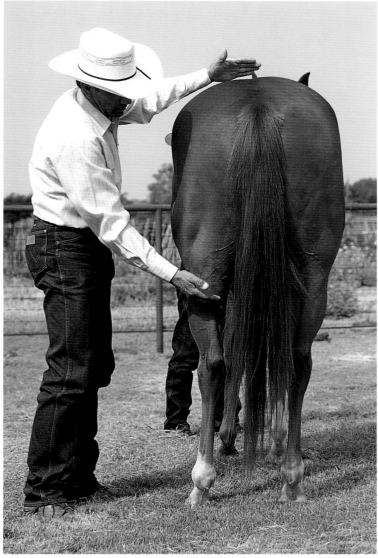

A barrel horse needs a lot of inside, hind-leg muscle on a long hip to be able to run, stop, and turn quickly.

underline handicaps a horse's ability as he has a 90 percent chance of overreaching (reaching with the back feet and clipping the front feet and legs) in the turn.

HINDQUARTERS

The hindquarters should be very long. They should not be flat on top, and should not slope (drop off) too much. There should also be a lot of inside, hind-leg muscle. These traits are necessary for a horse to run, stop and turn quickly. We like to see a double hip on a horse. From the side of the horse, a double hip looks like two humps on top of the hip, above the tail. If a horse has this double hip, he has more power in his rear end and can physically excel as a barrel horse.

A crease coming off the back of each side of the hip and continuing down toward the hock, with large muscling on both sides, is called a double muscle and denotes power and strength. If a horse is double-muscled and double-hipped, it has a lot more strength and consequently more athletic ability. Even though we like a horse with heavily muscled hindquarters, we don't want the muscle to be tight or bunchy, like a halter horse. Tight, bunchy-muscled horses sore more quickly and are choppy movers. We prefer long, smooth muscles.

The Wrights prefer long, smooth muscles rather than muscles which are tight and bunchy.

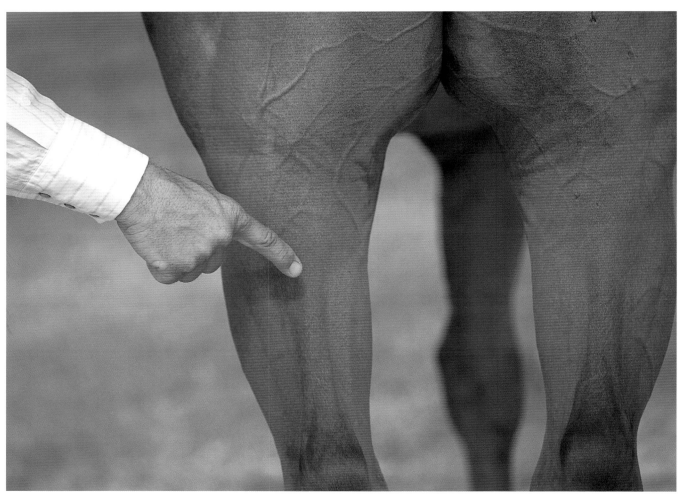

BACK LEGS

A barrel horse should be strong through the gaskins, with inside muscling being more important than outside muscling. The inside and outside gaskin muscles should have a lot of length and a lot of width for power when they have to push away from the barrel. Our personal theory is that good inside muscling is possibly more important than outside muscling. We feel that outside muscling controls motion, while inside muscling controls both motion and balance. A horse with a lot of inside muscle is athletic. Excessive outside muscle, like that seen on halter horses, is not necessary for a barrel horse.

There should be some width between the hind legs, both when the horse is standing and when he is traveling. Also, if a horse walks with his feet close together, he is not a good athlete. The reason for this is that where there is width, there is better balance. When the horse's feet are wide apart, his whole body frame has more support and better balance. If his frame is narrow at the bottom (his feet are close together), there is less foundation and support for the horse.

The width between the rear legs is the foundation of the movement of the horse. A narrow foundation, or structurally wrong alignment, causes poor athletic ability and a poor chance of the horse staying sound. Athletic ability is tied to soundness. If the horse uses himself wrong, there is stress on his soft and hard tissue (bones, muscle, ligaments, etc.), but with stress being placed improperly, it even adds more chance of injury.

A horse with his hocks too close together (cow-hocked) still has power, but it is a structural flaw. If a hock points out, however, it's even more of a structural flaw. The hocks should be as close to the ground as possible. Low hocks and short cannon bones belong together to create a horse that is athletic and stays sound. These attributes place the physical stress properly. As we described before, the longer the cannon bone, the more leverage there is against the upper portion of the body — shoulders, back and hips. A shorter cannon bone lessens the stress on these same areas and distributes it throughout the body.

The hind-leg muscle should start by the horse's tail and run all the way down to the hock. A horse's hind leg structure (hips, legs and feet) needs to be lined up, just like his front leg structure. Take that same string you used on the front legs and hold it at the corner of the point of the buttocks. The hind leg, hock, fetlock and foot should all line up when viewed from the rear.

A horse's hind legs shouldn't be dog-legged or sickle-hocked. With this type of weak conformation, the horse loses a lot of power and the ability to push out of a turn. Legs that sit out behind the horse or legs that are straight up and down are also weak and the horse is unable to get the propulsion he needs. Hocks that sit out behind the horse do not allow the horse to stop and turn quickly, because the hock must first be placed under the horse, and then perform the stop or turn. The hock that is built properly, allows an instant stop or turn. Also, low hocks that are straight are highly susceptible to injury.

When a horse moves, he should take his hind foot and stride over the track of his front foot. His movements should be long and ground covering. We like a horse to move like a cat — supple and soft with a gliding motion. We don't want a horse to hit the ground hard or have height and elevation in his knees and hocks when he picks up his feet. Excessive knee or hock action is not desirable in a barrel horse because there's wasted motion and time when the legs are in the air. Sometimes, it's a sign of soreness in the joint.

FEET

Anytime we are thinking of purchasing a horse, we'll start by looking at the horse's feet and go up from there. Most people prefer hoofs that are dark in color. It is believed that white feet are not as hard as black or dark hoofs. There's an old saying that if you like good feet, you don't want to buy a white-footed horse. But we don't buy colors of feet. We buy textures and some white feet have just as good texture as black feet.

The appearance and condition of the horse's hoofs are indicative of past nutrition, care and disease. A healthy foot is usually one that is proportioned well and has a good texture to it. A good-textured hoof appears waxy and is smooth and shiny. A good hoof has no cracks, has some moisture in it and is not dry and crumbly.

The coronary band should not be dry and leathery, but should appear slightly resilient. There should be no heat in the coronary band or the hoof. The frog should be firm and slightly pliable and appear intact, not ragged or spongy. There should be no odor or discharge from the frog, indicating disease, such as thrush. The sole should be firm and thick and not give excessively to pressure. The white line of the hoof wall should be free of discoloration and be uniform in thickness around the foot.

A horse's hoof should be round, not pointed or cut off square. The foot should be shaped similar to the coronary band, only larger. A big, flat hoof usually hits the ground hard. We like for the bottom of the hoof to be bigger than the coronary band in size. The slope of the hoof should match the slope of the pastern and the shoulder. This can be corrected, to a certain degree, by shoeing and trimming, but the hoof needs to have good balance naturally. The way the hoof is trimmed and shod, however, usually determines how healthy it stays. The heel should have some width to it and not be low to the ground. If the heel is too low, it will cause stress on the suspensory system because the tendons and ligaments will be pulled excessively as the foot lands. Also, the bulbs of the foot incur a lot of concussion with low heels.

A horse should not be mule-footed, meaning the wall should not be too steep all the way around the hoof. A clubfooted condition is also a cause for worry. With a club foot, the hoof wall bulges on one side, usually the outside.

SIZE

Even though the height of the horse is not the most important factor, we feel that a partially mature four-year-old should be from 14.2 to 15.2 hands tall to be the ideal size. Our goals for our prospects include the barrel futurities first, and a solid rodeo career later on. So a 16-hand horse, as a three and four-year-old, is usually like a 13-year-old kid who is six feet, four inches tall. He's constantly growing into himself and learning how to handle that big body.

There are some 16-hand four-year-olds that have great athletic ability, but most 16-hand horses are six years old before they are mature and can handle themselves physically and mentally. Consequently, we stay away from really large horses. Smaller horses, those around 14 hands, sometimes can't handle the ground when it gets deep. Since ground is a great factor in today's barrel events, you need a horse that not only has the ability, but also has the strength to handle really deep ground conditions. However, a great horse can run on all types of ground.

STRIDE

A horse's stride can make you or break you. A horse must have the ability to shorten and lengthen his stride. He must have a long stride to be able to get to a barrel, then shorten his stride to get around it quickly, and then lengthen it again every step he takes on the straightaway. To check a horse's stride, put him in a round pen that is at least 45 feet in diameter. If he pulls when he strides and uses his whole body as one complete mechanism, he will usually be coordinated. (Pulling means to place the feet on the ground and propel the body forward with a pulling motion.)

If he uses just one end, with his front feet churning and his back feet striding, then he is probably not built proportionately and will not be very coordinated. A churning motion is viewed as up-and-down, wasted productivity to create forward motion. Striding fluidly should create a forward motion that is moving smooth and fast, with the legs and feet moving front to back with little exaggerated elevation.

FLEXION

A horse must be supple and flexible in his body to be able to turn around the barrel the correct way. Some horses have too much flex, while others don't have enough. When you're leading or riding a horse that moves extremely stiff, from his ears to his tail, you'll have a hard time teaching him to wrap his body around a barrel. A supple, flexible horse should bend naturally when he is led or ridden. He should have the capability of bending or flexing either direction.

"PEOPLE NEED TO LEARN TO SPEAK THE LANGUAGE OF THE HORSE."

4

The Mind of a Horse

When Charmayne James and Scamper headed for the first barrel during the seventh go-round of the 1985 National Finals Rodeo, the pair were so intent on the job before them that neither realized how close to the wall of the alleyway they were. As Charmayne released the power of her horse, he scraped the wall, with the Chicago screw in his headstall falling out.

The great Scamper made the first two barrels with the bridle hanging from his head, but when he turned the third barrel after he had dropped the headstall completely off, the crowd roared with approval as the pair handily won the go-round, as well as the world championship, without the aid of a bridle. This can be credited to the training of Scamper, but we all know that the horse must have a mind and intelligence to accept that training and retain it.

The mind of a horse is very important in the training process of a barrel horse. Breeding and heredity have a lot to do with intelligence; however, it is also cultivated with education. A horse needs to be receptive to whatever you are asking him to do, but that's usually something that has to be developed. This can be done through hands-on education, starting the day the colt is born, and continuing throughout the horse's life. Some horses must be educated to learn and some are born to learn. This means that horses are not all created equal. They are born with different levels of intelligence, just like people.

Some horses are born smart and have a lot of try. You might think some horses are dumb because you haven't yet opened up their minds so they can focus their try in the right direction. To understand the mind of a horse, however, we need to remember that horses are prey animals that consider people predators, and each one thinks

(overleaf) Ed and Martha frequently break from their routine to ride their horses out in the pasture and give them relief from their daily training regimen.

differently. This is a critical distinction between a horse and a man. Prey animals are always afraid there are predators nearby and their first thought, if they are frightened, is flight. This is their method of survival. It is important that you make the horse understand that you are not a predator. You have to teach him to put Mother Nature aside, and sometimes that is difficult.

After you have accomplished that, as long as what you are doing is obvious and clear to the horse, he should understand. It may take longer with one horse than another; one horse may have more ability to comprehend or understand than another. But to be trainable, a colt should have the desire to communicate with a person and have the mental capacity to learn.

You can create friendship and communication with the horse by starting his education in the round pen, and from there progressing to the level of competition that the horse and rider are capable of achieving. Horses learn through repetition. We agree that in today's competitive activities, a horse has to think for himself, but only after he has had lots of perfect practice sessions, with you repeating over and over again what you are trying to teach him.

However, in some instances a dumb horse is better than a smart horse. Even though a dumb horse will wait for a rider's cues rather than doing something on his own, once a dumb horse has learned something, he doesn't usually figure out shortcuts or ways to do it easier. When a dumb horse is sore, tired or fatigued, he usually isn't smart enough to quit. He will go on and try for you. On the other hand, a highly sensitive, intelligent horse will learn fast, but if it makes him tired or sore to compete, he learns how to get out of it fast too.

A people-oriented horse, one that was handled properly as a foal, will be more receptive to training than one that wasn't handled, but that's not saying that a horse that wasn't handled early can't learn to be receptive. Just as with people, the earlier a horse begins training, the better.

You can teach a foal to use his mind. It is generally thought that horses have no logical reasoning power, unlike man. But they can and do think their way through situations if you allow them the time. Horses think like horses and people are usually the ones who need to be educated. People need to learn to speak the language of the horse. The language of the horse must be observed by the human to be understood.

When a horse is not understanding the human, he may show tension in several different ways, such as grinding his teeth, popping his tail, tension in his body or stomping his feet. The understanding horse looks and moves in a relaxed manner, has a soft look in his eye and reacts positively to the requests of the person handling him. For a horse to open up to you mentally, you have to gain his trust. With a fearful horse, you have to back up and start all over again until he understands what you are asking.

To determine what kind of a mind a colt has, start out in the round pen. You want a colt that is extremely receptive to your movement in that pen, one that really focuses on you. You want him curious about what you are doing and why you are in there. He should be very attentive as you move him around the pen. That means he's focused on you. He's always aware of where you are and what command you are giving, if he's been educated to that command. He should be responsive in a short time to the cues that you are giving him.

Move behind his hip and make a low-hand motion with your hand or a halter to make the horse move forward. Then step to his shoulder and be still, teaching him it's okay to slow down. When you step in front of him and, look him in the eye, blocking his motion, that should stop him and make him face you. However, for this last move you don't want your hands in motion while you're in front of the horse. The sooner he responds to these three commands, the more his mind is receptive to training. You can get inside a young horse's mind by just moving him around this way for a few minutes. You see what his responses are and evaluate him mentally. He can respond with his ears, tail and all parts of his body.

If a horse is thinking and receptive, his ears will move back and forth. A sign that a horse has a bad mind shows up in his lack of receptivity to training. His ears are usually set (locked in place, not soft and moving as if ready to act upon request). You can also tell by his tail. If he switches it

The position and movement of a horse's ears usually indicates his mood. When his ears are locked back, he's resisting and unrecptive. The surest way to gain his confidence is to educate him with a soft and easy manner.

every time you ask him to do something, that usually means he's not happy about doing what you are asking or that he is sore some place.

Try not to overload a horse's mind by overloading the memory bank and educating him to the limit. When that happens, the horse quits being receptive. For some horses, that's three hours, while for others, three minutes. Stop too early rather than too late! A horse will tell you with his facial expression and body posture when he's had enough. If you've been working the horse in the round pen for a while and he's still soft, he's receptive. If he's tense and jamming his feet into the ground, he's had enough. He may be standing dead still and start fretting and worrying. That means he needs relief. When a horse has had enough, you won't make any more progress; that's the time to quit.

When you work a horse into a negative position, you teach him to be negative. If you continue to the point of resistance, you teach him to be resistant. The same with competing. If you run your horse around the barrels three times a day, you can teach him to hate it. You need to be observant enough to know when you've reached the horse's limit.

A lot of horses that come from the race track have tremendous mental blocks because they have been handled differently from saddle horses. Many were broke to ride with blinkers on, so they've never seen a person get on their back. A horse like that has a definite fear of a person moving around behind his shoulder and on his back. Those types of fears are something that you have to overcome. Some race horses have the mentality to adapt very easily, but others need a major amount of re-educating.

COMMUNICATION WITH THE HORSE

Most communication with a horse is non-verbal. Our communication tools are our hands, arms, legs, feet, seat, voice and facial expressions. The golden rule in communicating with a horse is "to make undesirable things difficult for him and desirable things easy for him."

Some horses, like some people, are hermits. Their first impulse is to lock you out and resist you. Consequently, you must open their mind to receive education. You do that through communication with the horse. Your communication line with the horse is imperative for him to understand what you are asking. It forms a bond between the two of you. Horses aren't resistant if they are trained properly. The resistance comes from a bad line of communication and from a bad teacher. A good trainer will always work with the horse through cooperation and communication — not force.

Consistency is very important when training a horse as it promotes better communication between the horse and the rider. But the rider must also learn about the horse, just as the horse must learn about his rider. You must do everything the same way every time. That way, the horse understands what you are trying to communicate to him and what you want him to do. In competition, some riders ride their horses differently from the way they trained them. That really confuses the horse. Riders must be consistent in training and must also stay consistent in competition.

Also, the confidence of the horse must be gained through education that is soft and easy, no matter what the attitude of the horse is. It must be in a way that the horse understands on his level of thinking. The trainer must educate the horse in a way that not only fosters discipline but also gains the horse's confidence. If a horse is confident in you and what you are doing, he is more receptive to education.

When you first start a horse in the round pen, less is more, meaning when you get a positive reaction to your communication, in the early stages of training, quit right then. The next day, the horse will be more receptive to your education and react quickly, because he was rewarded by you stopping the day before.

GENDER

Even though we don't think a horse's gender (stallion, mare or gelding) makes a horse a better or worse barrel horse, the ideal horse for the average rider is a gelding. You don't have the problems that come with a mare or a stallion. You don't have to be careful where you stall him, what you haul him with or what you tie him next to.

Because he isn't distracted by hormones, a gelding can be more focused on his job. Most geldings are more dependable, honest and solid horses, especially if they haven't had any bad experiences. The disadvantage of hauling a gelding is that he can be hurt or crippled, and if he is permanently injured, there is nothing you can do with him except put him out to pasture and feed him. He can't reproduce.

The advantages of owning a mare seem to be that a mare's mental abilities develop sooner. Therefore, since they seem to be smarter, they learn quicker. If you find a mare that has the desire to try for you 100 percent, is easy to get along with and doesn't show when she's in heat, she will more than likely be a great horse. Also, if a mare gets hurt, it's not the end of her career. She can be bred and reproduce. Hormones can give a mare staying power, heart and try, but those hormones can also work against you.

When mares come into season (estrus), they tend to have varying mood swings. Those same hormones that gave her staying power can disrupt the rest of your barn or trailer when you are hauling down the road. Also, since a mare can be smart, you have to be careful that you don't make any mistakes in your training, because she will pick up on your mistakes as well as the things you want her to learn. Also, since they are smarter, mares tend to become bored faster. Sometimes, however, the fact that they are generally more sensitive than geldings makes them more difficult to ride.

If you are a novice rider and don't keep your mare focused properly, she might learn to out-think or cheat you — like turning the barrel before she gets to it, not

Because his mind is not always on what you're trying to teach him, a stallion will usually take longer to train than a mare or gelding.

making a pocket properly, not running home as hard as she can, or setting up before she comes to the finish line. Mares can sometimes figure the "easy way out."

Most people should not ride stallions, or even handle them. A stallion that is not educated and focused on his job has two things on his mind: protecting his territory and breeding something. You've got to keep him focused to train him and to haul him. He must always respect you and know where he stands with you at all times. Not many people know how to keep a stallion in that frame of mind.

Even if you are a good horseman and can handle a stallion, he'll usually take longer to train than a mare or a gelding, because he does not have his mind on what you are trying to teach them. Also, when you take a stallion away from home, new surroundings and new horses can cause even the best of stallions to lose concentration. That's one of the main reasons why stallions are hard to get ready for futurity competition.

However, if you do haul a stallion and have the ability and knowledge to handle him, he will usually have a lot of power, heart and try. Stallions seem to be smarter than geldings and have unique personalities. Some horsemen who know how to handle stallions get much closer to a stallion and prefer to ride them. There have been some women barrel racers who have hauled stallions, but not very many. Connie Combs Kirby was one, hauling Maudies Joak from coast to coast, winning the 1976 World Championship on him.

If you do haul a stallion, you must be sure to have a strong halter and a good rope on him at all times. Tie him good in the horse trailer and tie him carefully at a show. Also, you must be careful in the warm-up arena, where there are a lot of other horses around. Not only can you get your horse hurt, but you may also incur some liability for someone else or their horse, if you're not on your toes at all times.

"WE BELIEVE IN IMPRINTING ALL OF OUR NEWBORN FOALS."

5

The Colt's First Year

It takes a long time from that cool, spring morning, when a colt is born in our back pasture, to the first time he thunders down the alley in his first barrel futurity. Most barrel racers aren't willing to spend that much time, money or effort to raise a barrel horse. But the results can be gratifying, especially if you raise a big winner. But there are other reasons to raise your own barrel prospects.

Martha and I have found that it is better to communicate with a horse from the very beginning; therefore, we have decided to breed and raise some of our own colts. That doesn't mean that we won't buy some prospects that other people have raised, but whenever possible, we want to raise our own colts. We have a small band of mares that are bred the way we like them, and we have bred them to stallions that we feel will cross the best on them.

Previously, we bought three-year-olds that had already been to the race track. However, we found they were educated differently from the way we like them to be educated. They didn't understand our way of communication. So, several years ago, we changed our program by buying some broodmares and breeding them.

We believe in imprinting or desensitizing all of our newborn foals. We do this by touching the foal all over immediately after he is born. We put our fingers in his nose, mouth and ears and pat the bottoms of his feet, which simulates a hoof pick or farrier's hammer. We squeeze softly where the cinch will eventually go. We put our fingers in his mouth and ears and we may even turn on a pair of electric clippers and rub him on his sides, but we don't clip him. We may even take a plastic trash bag and get him used to a plastic bag touching him and making strange noises.

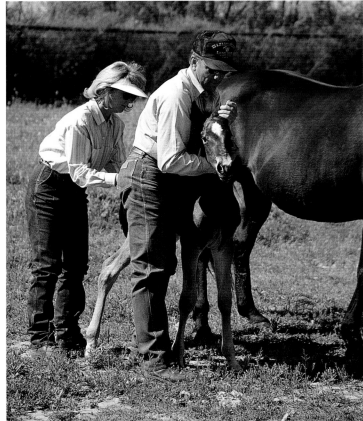

We desensitize every part of the colt's body except his rib cage, because that is where we want him active when we touch him with our legs later on in life. By not desensitizing that area, it will remain sensitive to leg cues. We like a colt to be familiar with us from day one. But, that doesn't mean that he should be like a pet; we just want him to be comfortable with humans and "bonded" to us. Bonding means that the colt trusts us and that there is a clear line of communication between us.

Throughout his first 12 months of life, he's never just turned out and left alone to grow. We get our colts up regularly and work with them. If the only time we got a colt up was to give him shots or worm him, the colt would form negative feelings about us. Therefore, throughout the first year of his life, we try to handle each colt and educate him periodically. When we get a colt up, we start him in the round pen with the Tom Dorrance method of resistance-free education. This is basically the same educational process which was discussed in Chapter 3, when we were analyzing the athletic and mental qualities of a young horse.

When we walk into a pen with a colt, we don't want him to turn his rear end toward us or turn and leave. When we turn the colt loose, we teach him to react to our body signals. The object is to have the colt face us and preferably come to us when we ask him to. We should be able to handle him while he's loose, with him understanding our body signals. He needs to come to us to be caught, and he needs to be ready and eager to do what we want him to do.

When we step to his rear and ask him to move, he should move forward. If he doesn't, we move our hands or even swish a halter at him. Our body language should also tell him that when we move to his shoulder, he should slow down. If he slows down, or even gives us an eye or an ear, we yield by backing up. That teaches a colt that if he does what we want him to do, we give him some relief. He should also learn fairly quickly that when we step to his eye, he should stop and look at us.

If we try to touch or catch the colt, and he leaves us, we propel him forward with our hands or a halter. Make the colt work by moving him around the pen at a good clip. However, as soon as he gives us any ear or eye indication (he'll look toward us or flick an ear our way) that he's wanting to pay attention to us, we give him room by backing up. We let him focus on us. It won't be long before he'll let us catch him. Horses are natural followers and they are looking for someone to lead them from the day they are born.

If he turns toward us and lets us catch him, we reward him when we walk up to him. We tell him how good he did in a manner that the horse understands, such as using soft words and gentle rubbing with our hand. Horses relate to the tone and the smoothness of a human's voice. They are much more aware of people and are better judges of character than people think. When you're rough and brash around a horse, he reads you that way and doesn't want to come visit you.

(previous pages and opposite) Imprinting is an important part of a colt's first year. The Wrights desensitize every part of the colt's body, cleaning his nostrils, scratching and rubbing his body, and fooling with his tail and ears. They won't desensitize his rib cage which will remain sensitive to leg cues when they begin to train him.

After working a two-year-old in the round pen, Lonny Johnson greets the colt with a soft rope, and rubs the rope on his face and body until he gets used to it.

After rubbing his body with the soft rope, put the rope around the colt's neck and pull him in both directions until he yields to the pressure.

We halter a colt in the round pen and get him so that he not only leads with the halter, but we also can stop him, turn him both ways and back him up with the halter or a soft rope around his neck. We get him to yield to soft pressure. That means when we pull the head or neck gently to the left, and he yields one or two inches, we release right then. That teaches him that when there's pressure and he yields to it, he instantly gets the reward of no more pressure.

Later, when we move out of the round pen to a larger pen, we use the same communication. Anytime he comes to us, we're soft with him and never drive him away. The only time we will drive a colt away is if we are horse back. That way, he won't associate us with being driven away.

We try to work with each one of our colts weekly because a colt's mind needs to be refreshed often. A young horse is exactly like a child in school. If he gets to goof off too much, he doesn't retain information. He needs to be reeducated within a short period of time. We also fool with a colt's legs so that later we can trim his feet. When we handle his legs, he soon understands that when he yields, we yield, and give them back.

Part of the time we work the colts together as a group and part of the time we work them one at a time. We work them together so they learn they should stand still while we walk up and catch one and lead it out of the bunch. We don't want the others to break and leave, by having fun and playing. However, colts are much harder to work with as a group, so you need to have some experience to work with them as a group.

Bonding takes different lengths of time for colts with different personalities and levels of education. For some colts, it takes a lot of thought from the educator. They truly lock you out. Others, when you show them some relief, hook up with you and bond really quickly. Still, with others, you have to find the key to them. With some colts, if you back clear across the pen, and they give you an eye, that's the first sign of bonding. You have to decide if that's enough for that day and quit, because they may be the kind that are so secluded in their minds that if you put more pressure on them, you send them away from you. With other colts, the first day they yield to you, you can get them hooked up to you, and when you walk off, they're glued to you like a magnet. Each colt is different.

If we have a certain colt that is a little more distant than the rest and really doesn't care about being fooled with, we put him in a stall for about 10 days, away from the other horses. That way, we're his only friends that show up to feed and water him and let him out of that stall. He'll soon learn to do things with us and will become bonded.

After you have fooled with colts out of certain mares, you are much more aware of which ones will be easy and which ones will be tough. Some kind and sweet mares produce distant colts, until they bond with you, and then they are great friends. We have a colt that is unfriendly, but his mother taught him to be like that. She had been mistreated by someone when she was a foal, so when you walk in the pasture with her and her colt, she will leave with her colt. She taught him not to trust people, and he's been difficult to change.

HANDLING A COLT'S FEET

You should be able to handle a colt's feet from the day he stands and nurses. We're not saying that we actually trim a foal's feet that young, but by going through the motions, we are helping to educate him to the whole process early in life. When the time comes to trim and eventually shoe him, he is comfortable with the procedure.

The growth and development of a foal's legs are affected by the levelness of his hoofs. Therefore, when a foal is 30-60 days old, his feet should be trimmed level for his joints to develop properly. Also, you should keep the toe from growing out too long as long toes cause stress on the suspensory system, especially on the back of the legs. When the suspensory ligaments get sore, the other parts of the horse's body also get sore. For further information on trimming feet, see Chapter 15.

HEALTH CARE

It is important that the health of your colt be maintained and monitored. He should be on a regular schedule for shots and deworming. See Chapter 14 for a health care schedule.

Use common sense, and work with a veterinarian or equine nutritionist, when you decide on a feeding program for a young horse. A colt needs more protein while he's growing, up until his three-year-old year. However, you'll probably be worse off feeding too much protein than too little, as it can cause the growth plates in his joints to develop too quickly. There must be a balance in everything that you do. There is a fine line between doing too much and too little. For suggestions on feeding, see Chapter 14.

By raising a colt "your way," you have control of all his early experiences, which is very important to things that he does later in life. But again, remember, you must be dedicated to this endeavor as it not only takes a different kind of knowledge from training horses and running barrels, and it also takes time, money and proper facilities. The results, however, are worth it!

"HE CAN'T LEARN IF HE IS IN PAIN."

6

Starting the Prospect

If you have just started barrel racing, your best bet is to purchase a solid barrel horse that can teach you the ropes. That way, you can get the feel of what it's like to rate a barrel. You can find out how it feels to skim a barrel and, with a sudden burst of speed, head for the next one. And you can learn about speed and how you react to it.

But after you have learned how to ride a barrel horse and how to win, the next step is to train a horse of your own ... from the beginning. However, before you can begin training a young prospect, there are several considerations to take into account, including age and physical and mental maturity of the horse.

Working colts on barrels is Martha's speciality and starting the colts is my department. I try to find outside help to start my colts. However, since it is very important that they are started the right way, I usually end up doing most of the initial ground work, as well as round pen work, myself.

I rarely saddle a horse before he is two years old, because I feel that before that age, he is not physically mature enough to carry weight. A colt is obviously not ready to ride when he is too small or when his bones haven't developed enough. But maturity doesn't mean just size. Size is important though, because I want a colt to be over 900 pounds and stand over 14.1 hands before I start riding him. But muscle definition and how the colt handles himself (whether or not he is clumsy and awkward) is also important.

Even though the colt may have good muscle definition, height and bulk in the body, he must also have a certain maturity in his coordination. I don't want to start riding a colt that is still in the clumsy, awkward stage. When a colt works properly in the round pen, moving smoothly and collected, handling himself with some maturity, that's when

(overleaf) If a colt bucks when he's let loose in the round pen, step behind him and pressure him to go faster. When he lets up, step back and he'll soon learn that bucking is not good for him.

he's ready to start riding. However, if I have a horse that is still growing and going through a "leggy" stage, he can still be trained; I just can't make him overly tired, because a horse that is fatigued can injure himself.

Also, most young horses, especially two-year-olds, have "open" knees, meaning their knee joints have not completely formed. This can be seen when looking at a horse from the side, as his carpal (knee) joints will be irregular, giving the impression that the carpal joints are not fully closed. This is often caused by a mineral imbalance. As the horse matures, the joints usually become more pleasing in appearance. However, sometimes knees can look like they are completely formed and they aren't. Therefore, the only way to really check is to have the knees x-rayed.

Mental maturity is as important as physical maturity. A colt is mentally mature when he can retain a lot of groundwork in and out of the round pen and when he can stay focused. Some can be focused as yearlings, while others can be four years old before you can hold their focus. Mental maturity is an individual trait that is usually hereditary, but also comes from a background of good education. There are certain breeds, such as Thoroughbreds, that mature physically early on, but are late bloomers mentally. Therefore, horses with a lot of Thoroughbred blood in them are often not good prospects for barrel horses.

The sign of a mentally mature horse is when you can communicate to him that you want him to go forward, stop, or turn around, or that you want to pick up his foot. A horse that is mentally unprepared for training, is one which doesn't focus on you when you are working with him. Also, he doesn't retain what you have taught him the day or week before and he is flighty in response to your communications, especially when you are starting to change things. An immature horse has negative reactions to your communication.

Actually, I can make decisions on the mental maturity of a colt when he is still a weanling and I'm working him in the round pen. It takes the experience of working with a lot of young horses before a person will be able to make this type of decision on a young prospect.

A colt needs to be educated first in a round pen. The method that I use to break our colts is the Tom Dorrance method. This is the same method used by Ray Hunt and many other modern-day horsemanship clinicians who train horses by getting into their minds rather than by using force. They communicate with horses, not intimidate them. They do this by developing lines of communication that both parties understand — human and equine.

Any time I start working with a colt, my goal is to open a clear line of communication. I want him to comprehend what I ask of him. At the same time, I need to clearly understand his responses to my requests. In other words, when something is bothering the colt or he is tired or sore, I should recognize those signs and act accordingly. I get off his back if he's too tired to continue and tend to the places where he hurts. He can't learn if he is in pain.

Young horses, especially, have short attention spans. If the young horse you're riding is doing the things you want him to do properly, but you're losing his attention span, quit for that day. With a colt that's still fresh, feeling good and having fun, you can go farther.

HOBBLES

Before saddling a colt for the first time, I use hobbles as a training aid if the colt moves or shows signs of not standing still to be saddled. Hobbles are a piece of equipment that go around each of a horse's front legs to restrain him. They help to teach a horse manners, discipline and patience.

Hobbles should be made of soft, wide straps that fit between the pastern and the coronary band of the horse's feet. There are many types of hobbles on the market, mostly made of leather, fleece-lined nylon or rope. I prefer leather hobbles that are connected by a chain. Outfitters commonly use them on their pack animals. I don't think all-leather hobbles are strong enough. Also, they will sometimes slide back and forth, drawing down on the pastern. I don't like anything that is tight on the pastern as I want to determine how tight the hobbles are. I don't want them to restrict the area around the pastern; however, they shouldn't be loose enough to come off over the hoof either.

Hobbles will help teach a horse manners, discipline, and patience.

Even though most horses adjust to hobbles rather quickly, they still need to be hobbled in a confined area with soft ground in case they fall. However, an experienced horseman should work with the person hobbling the horse to help protect both the horse and the person. Different horses react differently to hobbles, and you can never be sure of what is going to happen.

A horse which has never before been hobbled can be frightened by having his legs restricted. A horse's two natural defenses are to run away or to kick at whatever scares him. When putting hobbles on a horse for the first time, it can be dangerous. A scared horse can kick or paw you or even fall on you. If you are a novice horseman, seek the help of someone experienced in hobble training, rather than trying to do it yourself.

After I have haltered a colt in the round pen, and he totally understands bonding and yielding to pressure, then I am ready to start educating him to hobbles. With a halter on the colt, I put a very large, soft rope under the fetlock of one leg. Holding both ends of the rope, I pick the leg up and get the colt to yield to the pressure of the rope. When he yields, I reward him by yielding back. Yield from the animal, followed by yield from the educator, forms a bond and makes the animal quiet and secure.

When the horse yields to pressure on every leg (which usually takes several days), then I hobble him in a round pen with a halter on him. I never hobble a horse and turn him loose to learn for himself. There is a good chance of him injuring himself and worse than that, it will damage the bonding and faith he has in me, because through my hobbling him, he has been allowed to hurt himself. The pain is related to me because I hobbled him.

Just like people, each horse has a different attention span and speed for learning. After six sessions, the average horse will yield to hobbles. These sessions should be within a two-week span, because a young horse has a short memory and things need to be repeated within a short span of time. After he has matured and been handled more, it can be a little longer between lessons. But a horse is an animal which learns by repetition. There has to be perfect practice and repetition for the horse to understand. Just randomly doing something does not educate a horse.

You should also start in the round pen if you want to hobble an older horse. If a horse hasn't been educated properly, or is not educated at all, to something like hobbles, you need to lay a foundation for that horse to understand your communication. Even if he's 10 years old, treat him like he's four months old and go through the same steps as you do a colt. Once your colt is used to hobbles, it's a good idea to hobble him after you haul him away from home for the first time. You should also hobble him when you saddle him and while you have him tied up. This keeps him from walking around and pawing the ground. It teaches him manners.

THE FIRST SADDLING

Before I saddle a colt, I reinforce our "body language" drills in the round pen. I step to his shoulder and if he wants to leave, I let him, but keep him circling the round pen. The idea here is that if he won't stand still for me to approach him, I make him work. I use body language to make him go forward, slow down and to stop. If he doesn't move forward when I step to his hip, I may move my hands or even cluck to him to make him go.

To encourage him to slow down, I move to his front shoulder and if he slows down, I back up, giving him relief. To teach a colt to stop, I move toward his head, effectively blocking his forward motion. But I don't step right in his path and take a chance of getting run over. I just cut him off by decreasing the space he has to run in. When he perceives that he's out of space in that direction, he'll stop and usually turn back the other way.

Even though clucking and the word "whoa" are important for a horse to know and respond to, I don't do a lot of verbal cues. This is because later on, when you are running a horse in a barrel race, it is real hard to give verbal cues due to other noises, like announcers and a cheering crowd, that make it impossible for him to hear me. But I am the only one that is physically cuing him.

When he has been worked down, is listening and responding to my cues of "go forward," "slow down" and "stop," it is now time to saddle him. For some horses, this preliminary work may take minutes, while for others, it may take an hour or so. It all depends on how open the horse's mind is to education.

The first saddling takes place in the round pen, just like his other education. I don't bridle him; I have only a halter and a long lead rope on him — not a longe-line, but a soft, lead rope that is at least 16-feet long. A soft rope is much safer for your hands and the horse, especially if you want to use it to reach down and pick up one of his legs. A hard, nylon longe-line strap can burn your hands if the colt decides to take off and run the other direction.

The colt must first become familiar with the saddle pad, so I walk to the front of him with the pad. I don't intimidate him or charge him with the pad, but I just take it to him and let him smell it. Then, I start rubbing his body with the pad, which, in cowboy terminology, is sometimes called "sacking" the horse out. I never slap him with the pad nor do I make quick moves that he doesn't understand. I rub the saddle pad everywhere on the horse's body — his back, neck, shoulders, ears, tail, legs, under his belly and on both the right and left side. Always work equally on both sides with any training methods.

If the horse is distant, arrogant, tough-minded, or acts like he wants to be a bronc, it might take a couple of weeks before he gets used to the pad. However, an intelligent horse, that is also real quiet, might accept it the first day. And by the second day, he will be ready to go on with the saddling process.

After the colt is used to the saddle pad, I do the same thing with the saddle. I let him smell it and then, very gently, I ease it over his back, withers, neck and hips. I rub the saddle all over his back, from back to front and from front to back. Even though the saddle won't eventually be anywhere but on his back, I still want the colt to be familiar with the saddle on all parts of his body. Someday it may slip and I don't want him scared of it.

During the time I'm doing all these things with the saddle, the colt should be focused on me and he shouldn't be looking outside the pen. Anytime I want to regain his attention, I do what the colt's mother would do. She'd take her shoulder or hip and bump him. She might even reach over and bite him, but not hard, as it's not intended to hurt him. It's just to get his attention.

I don't bite the horse, but when I'm trying to get his attention, I do bump him or just push him a little off balance. That will make him look at me and, as long as he's focused on me, I talk to him and brush him with my hand. Even

Lonny Johnson "sacks out" a colt before saddling him for the first time, rubbing his body with the saddle pad until he becomes familiar with strange, new objects on his back.

Allow a colt to smell the saddle, then gently ease it over his back, withers, neck, and hips until he is familiar with it on all parts of his body. Cinch up the saddle just tight enough so it won't slip. Walk him around and then cinch him some more. Never cinch up the saddle tight right away. That's inviting trouble.

when I'm saddling him, I stroke him on the neck. You can do things somewhat quickly around a horse as long as you do it smoothly and he's aware of your next move. I never put a saddle on a horse and cinch it up tightly. I always pull the cinch up lightly, just enough so that the saddle won't turn when the horse moves around. Then I walk the horse around and cinch him up some more. I repeat that until I have the cinch as tight as I want it.

When I move the horse around, I'm giving his skin and hair a chance to smooth out under the cinch, blanket and saddle. Also, I'm allowing his muscles to become accustomed to the pressure before I tighten the cinch. This cinching-up procedure goes for any horse — an older, educated horse, or a young one.

After the colt is saddled, I turn him loose, with nothing on his head, and get him to travel around the round pen. I control him with my body during this exercise, just as I did before I saddled him, walking toward his hip to get him to move; at his shoulder to make him slow down, or toward his front end to get him to stop.

If the colt bucks, I get behind him and put a lot of pressure on him, making him go forward faster. This teaches him that bucking is not wanted. The instant he lets up, I step to his head, then back up and yield to him. This positive move shows him that it is good for him to quit bucking.

MOUNTING THE COLT

Even though some colts can be mounted the first day they are saddled, I usually like to take more time than that. I think getting them used to the saddle is enough for a one-day session. I've taught this colt how to move with a saddle on, how to turn around and how to stop.

Before I get into the saddle, the horse should have a halter and lead rope on his head. I take him to the center of the round pen, move the stirrups up and down and back and forth, pat the saddle and then slap it softly. I do this on both sides of the colt, until he is totally comfortable with the saddle and the noises it makes.

If he's a really trusting, highly intelligent horse, then I might get on him the first day that I saddle him. But it's a little better just to desensitize him to the saddle and get him to travel with it. Then leave him alone. To desensitize a colt to a saddle, move it all over him, jerk on the stirrups and throw them up on the saddle, getting him used to anything you might do with the saddle. In the early stages, less is more. The less you teach a colt in one day, the more he will learn. If you try to cram a lot into his mind in one day, you'll get fewer results.

Before I mount the colt, I bump him softly with my chest, right where I would be getting on him. I jump up and down on the ground, hitting the horse softly with my chest. This gets the horse familiar with motion in that area. Then I stand in the left stirrup, but I don't throw my leg over the saddle. I just step up on his side, totally against him, and reach across him with my arm and chin, just

After the colt is used to the saddle, Lonny steps into the left stirrup and stands there for a moment. If the colt shows no resistance or desire to move, Lonny swings his leg over and sits in the saddle.

barely above the seat of the saddle. This teaches the horse to hold the rider's weight and shows him how it feels for the rider to be in that position. If he decides to step forward, I have the lead shank where I can bring him back around me. Hopefully, he has enough trust in me to just stand there. I reach across and pet him on the other side, starting at the neck and working my way back to the shoulder, rib cage, flank and hip.

If the colt moves when I begin to mount him, he needs to be taught to stand still. When he moves (while I'm still on the ground), I pull him around me with the halter rope. I step up against the stirrup and keep him rotating around me until it is his idea to stop. I keep soft pressure on the rope and keep him coming around me. When it's his idea to stop, I release the pressure on the rope.

Then I step into the stirrup. If he's moving, I bring the rope around to my leg in the stirrup and keep him circling me until he stops. Then I release his head and, still standing in the stirrup, I reach across him and pet him on the hip and shoulder, telling him "that's a good idea." I don't tell him "whoa" during this time, I just wait until he stops on his own. Since I have hold of the lead rope, all he can do is move in a circle around me. He will soon tire of that and stand still, because that's easier.

While he's standing still, I throw my leg over, sit in the saddle and pet him, rewarding him for letting me on while his feet are standing still. When his feet and body are still, I'm controlling his mind. I step up on both his left side and his right side because I want the horse to be ambidextrous. He needs to get used to a rider being on both sides.

When I do move my leg all the way over and sit in the saddle, I don't expect the colt to move off immediately. I get him totally accustomed to my weight and motion, the sight of a rider on his back and the noise of the rider sitting in the saddle. Again, I don't use a lot of verbal commands with a horse. I use motion, visibility and feel as cues.

The next thing I want the colt to do is travel forward. With the halter on his head and the lead rope in my hand, I can now move him around the round pen. I never pull the rope for any reason. I leave his head totally free. He'll understand the saddle. And if I have communicated properly, he will understand that when I step up onto his back, I'm not a foreign enemy being on his back. I'm a friend, and all I want him to do is travel forward.

I never kick or hit him; I use my "body throttle" to move him forward which means that I educate him so that when I'm in the front of the saddle, up against the swells, he should move forward. When I'm in the front of the saddle, giving him a loose rein, that's his incentive to go forward. I may even slap my leg, the fenders of the saddle or the saddle cantle as an incentive to go forward. The horse usually relates to that and moves off a step, and when he does, I know I've communicated with him.

I teach a horse to slow down when I'm in the middle of the saddle. I also pick up my hand and put light pressure on his face by softly tugging on the rope I have tied to his halter. When I sit in the back, I totally stop the horse with a light rein or rope pressure. Pretty soon he has learned that when I sit down in the middle of the saddle, he is supposed to slow down or shorten his stride. When I'm in the back of the saddle, I want him to stop.

Once we are traveling forward, I can take the rope that is attached to his halter and, with my hand, elevate it. I move it in the direction I want him to turn. By turn, I mean turn only two or three inches. The instant that I get response from a young horse, I give him relief. When I do give him relief, if he wants to stand there and digest the thought of what we just did, that's great. Normally a colt does.

Then I flip the rope in front of his face to the other side. I do the same motions, this time going in the other direction. By flipping the rope in front of the horse's face, it not only puts the rope on the other side, but it desensitizes him to the motion of the rope. I don't want to spook him. I do it softly and gently.

If the horse finally goes forward, but then doesn't want to stop, I let him travel until it is his idea to stop. That usually takes only a few minutes and not more than five or six minutes on a young horse, as they are not used to carrying weight and tire easily. When he tires, he'll think stop. More than likely he will stop if you have brought this colt up with the proper education.

In the round pen, when you were on the ground, you taught him to stop from nose pressure on the halter; turn from side nose pressure and come to the rope that is picked up. A colt should already be educated to stop and turn around as a baby, so you shouldn't have any trouble when you get on him.

In other words, if the horse has had the proper ground work leading up to my getting on him, I never have a problem. If I cinch a horse too tightly for the first time, I can have a problem. But if I am always in control — and there are no outside factors, like a horse running by the round pen or a feed sack blowing through the round pen, breaking the colt's focus — I shouldn't have any problems.

The colt could buck the first time I mount him, and if he does, I never touch his head. I let him buck until it is his idea to quit, then I pick his head up. I realize that not everyone can ride a bucking horse and this usually takes an experienced horseman. But it's very important to stay on, because if the horse does buck you off, you are teaching him how to get rid of you.

The amount of time that I would spend with a colt in the round pen would range from 15 minutes to an hour per day, because by that time, fatigue would set in. And fatigue works against you. Fatigue borders on intimidation training. Just remember, this all doesn't happen the first day, even though it can. It all depends on the horse. Let the horse tell you when he's ready to advance to the next stage.

The first day that I get on a colt in the round pen, all I want out of him is forward motion and stop. The next day, if the communication is thorough, I teach the colt to go, stop, and turn around just a little bit. Then, I barely increase it until I have the colt so he will go, stop, turn, and when he stops, he can back up one or two steps. This can take a week or two. Then, I turn him out and give him the reward of three or four days off.

BITTING WITH A SNAFFLE

After two or three days of riding the colt with a halter and lead shank, he should be ready for a ring snaffle. Some people use a rope noseband or a sidepull at this point, but in my program, I feel a horse is ready for a snaffle the day that I'm going to hang something on his head. I don't like to use anything else because the colt still has to make the transition to the bit.

I put a ring snaffle in the colt's mouth for five or six hours per day for two days in a row, just to get him used to carrying it. The ring snaffle should be adjusted so that the colt lightly grins on each side of his mouth. It should never look like he's laughing, but on the other hand, the ring snaffle should never be loose and hanging in his mouth. If he's slightly grinning, he will get used to a little bit of cheek pressure. If you get on him with the bit too loose and it comes up and touches his cheeks, it will surprise and scare him, as he's not used to that feel. Since the colt has been educated to the halter and lead shank, I put the ring snaffle on his face, with the halter still on his head. I get the horse to come to the same pressure, handling the ring snaffle lightly and simultaneously with the halter. Since the colt's nose is already educated, that helps educate him to the ring snaffle.

On the first day I use a ring snaffle, the only way I turn a colt around is by extending my hand in the direction I want him to go, which puts a little pressure on the colt's tongue, bars and cheek. I never increase the pressure. Just pick the rein up, about saddle horn high, and hold the pressure until the colt yields from two to four inches. The minute the colt gives a two to four-inch yield, I release him immediately, which is his reward. He will soon understand, that if he feels pressure and yields to it, he will get relief. That's what educates a horse and makes him light on the bit.

The colt has already learned to yield to hand pressure, since he has been yielding to it in a halter since he was a baby. Therefore, by this time the horse knows how to follow his nose, by yielding to pressure. I now use a little "body English," like some leg pressure from the calf of my leg in the horse's flank or rib cage, but I never use leg pressure in front of the front cinch. I want the horse to learn to move away from pressure behind the cinch, but not in front of it.

When a horse is moving properly and I want to turn to the right, using the lead rope and leg pressure, I take my right leg and barely move the rib, shoulder and hip over to the left (about three or four inches). My hand movement on the lead rope is followed by leg pressure. Then I release that pressure, because when I touch him with my right leg, it puts the horse's body in position to turn right and the pressure should cause him to put his head to the right, looking at the pressure. A horse never throws his head away from pressure, unless that is educated into him.

Then, once the horse has started in the right-hand mode to turn, I release that pressure, and follow it up with light outside (left) leg pressure. Both pressures are dictated by the reaction from the horse. When I get reaction from the horse, I take the pressure off. When he moves his rib, shoulder and hip over and is in position to turn right (which means moving it over and to the left just slightly), then I use outside leg pressure. As soon as he yields, I quit the outside pressure. The reason I use inside leg pressure first is not only to put bend in the horse, but also because it's an easier physical position for him to stop flat and turn flat.

In the days ahead, I want two things to happen through my teaching a colt to turn in this manner. I want pressure from my inside calf to build "flex" and "side-pass" position for a barrel; but I want my outside leg pressure to push the horse up into the bridle and get his body to follow his nose around the turn. I don't push a horse around the turn with outside leg pressure, but if a horse ever lets his shoulder or rear end drift out, I contain him with outside leg pressure.

A horse thinks with one side of his brain at a time. When I educate the right side to right leg pressure and the left side to left leg pressure, I want a horse to go forward when I touch both sides. When training a colt to stop, I use only one side of his mouth at one time. I make contact with the horse's mouth by lightly touching the reins on the right side three or four times. This should bring his right hind leg under him. "Touching the reins" means my hands have contact with the mouth and then I release that contact. When I "touch" with the rein, I have contact with the mouth, whether it be a tight rein or just slack enough to cause contact on the corner of the mouth; or in the case of a lead shank, I have contact with the horse's nose with a halter.

To get a reaction from the horse, this contact can be done with the fingers, but your wrist must enter into it, along with a slight arm motion. The wrist and fingers are sensitive and reaction from the horse comes fast. The whole arm is strong; however, it takes longer and the reaction from the horse comes slower. The wrist and fingers are 90 percent of the action, while the whole arm is approximately 10 percent of the action.

When stopping, I don't use voice commands. As I mentioned before, voice commands can get you in trouble later. What I am trying to do here is not a complete stop. I'm just slowing the horse down by forcing him to put his hind leg under him. Then I touch the left side of his mouth three or four times lightly, getting the left hind leg to come under him. When he understands that a touch on the side of the mouth means to bring that hind leg under him, I gently seesaw my hands back and forth. This cue should help him to understand that I want him to put both back legs underneath him.

After teaching a colt to stop, I ask him to take one step back. Backing up is an unnatural motion for a horse so I don't keep drilling that into him. I do it once and the maybe a half hour later I do it again. It usually takes several days of stopping before the colt learns to back up. I teach one side to back up, then the other side and I may even use some leg signals with my back up. But I don't add leg signal when I first teach a horse to stop, because up until then, the colt has been trained to go forward with leg signal. Once the colt starts backing up, I add the leg signal to increase the speed of the backup. I never use a solid pull to back up.

All of this usually takes from two to five days, or it can take as long as two weeks, before the colt has digested what I have taught him. The colt can be in a snaffle bit before I leave the round pen, but it's not necessary. Some colts do fine in a halter if their nose is educated to that halter, yielding to pressure. As soon as I can handle the colt by turning and stopping him, I will take him out of the round pen and into the pasture. By that time, however, I have the colt in a snaffle bit.

When riding the colt in the pasture, I travel quietly and easily. I always start out at a walk and do not move to the trot until he does everything perfectly at a walk. After I

After a few weeks of riding, a colt should know that when the rider moves forward with his reins and body, he should also move forward.

The colt should soon learn that when the rider sits down, picks up the reins and sits back in the saddle, it means "whoa." After stopping the colt, back him up to reinforce the "whoa."

have ridden for 15-20 minutes, I gently turn the horse to the left. Then I'll travel a little farther and gently turn the colt to the right. I repeat this until the horse learns and turns properly. Then I'll stop, back up one or two steps, and then quit for the day.

After about two weeks of riding in the round pen and then moving to the pasture, I turn the colt out from six to 10 days and never touch him. Then I bring him back up. The colt should retain a high percentage of what he has learned. However, anytime there is a problem, I go back a step or two to the foundation and refresh his memory.

After a colt has been started for several weeks, I want him to understand why I catch him, why I saddle him and what I expect of him after that, including traveling, moving forward quietly and easily. He should know that when I move forward with the reins and my body, he should move forward. Sometimes I create energy by slapping my leg or tapping him on the rear end, or moving my legs gently in the air (not kicking) just outside his rib cage, to urge him forward.

He should also know that when I sit down, pick up the reins and sit back in my saddle, that means "whoa." Anytime I start a colt, I'm very precise, definite and repetitious with my cues so the colt clearly understands the cue the same way each time.

TIME IN THE SADDLE

How long I ride a colt each day is dictated by his advancement. It is important in the early stages of training that I get a colt soft and relaxed in the round pen first before I ride him out in the pasture. This usually takes from three days to two weeks. In the round pen, a horse can learn to buck or be hyper or aggressive if handled incorrectly, so I work him each day until he is a little bit fatigued. However, that doesn't mean that I need to have him totally fatigued, where he's not receptive to education. There's a fine line between having a colt calm in the round pen and having him "give out." When I get the results I want in the round pen, I quit for the day. You can work a good-working horse into a bad-working horse by working him too long.

Once the colt has advanced enough that I can ride him in the pasture, I no longer ride him in the round pen. This means that he is soft and relaxed in the round pen and still soft and relaxed when I take him to the pasture. Unless I have some kind of problem getting on him or some other problems occur from not properly laying a good foundation, from then on, I ride him in the pasture.

A horse can be ridden in the pasture from three to six days a week. I skip every other day for awhile, then I come back and ride him again. Some days, I just pleasure ride him. I don't put the colt through a training session every time I ride him. I don't want him to associate me with work only so that he begins to resent both the work and me.

Once the colt fulfills the goals that I want early in his two-year-old year, which are traveling with me aboard, turning around, stopping and backing up, I turn him out and let him grow. In Texas, we usually turn the colts out during the hot summer months, as it is usually too hot to ride a two-year-old any way. After I've turned a horse out for a month or two, I get him up and ride him for a few more days. Then he is turned out again for a month or two. This is done to help him develop physically and mentally, so that when I do ride him again, he will be stronger and will think a little more.

It is typical for me to ride our colts about two weeks and then turn them out for a month or so. Ideally, I will get a colt up and ride him for two or three days every month. If he really learns fast, I get off of him. If I ask too much of him, he might give up. The colt's mind could probably take it; however, it could take the heart and try out of him if I worked him too hard. Also, his body can't take too much work at that young age.

PICKING UP THE PROPER LEAD

In the fall of his two-year-old year, I feel it's time to lope circles and educate the colt to picking up the correct leads. Proper leads are imperative because it is necessary that the horse be in the proper lead before he turns each barrel. It is also necessary that he be in the proper lead when he leaves and drives away from the barrel. If your horse doesn't take his proper leads while running, or cross leads (being on the correct lead in front but on the wrong lead behind), he can't perform as well and he leaves himself open to injury.

When starting a colt, I don't over emphasize leads. To eliminate stress, I start out in a large circle, about 90 feet across. After he has trotted the circle several times, I take his head gently to the outside of the circle, laying a leg on the outside flank and putting weight on the inside stirrup, while urging him to lope with my outside leg. By "urging" him, I mean rolling my foot or spur up his side or rib cage. This is soft communication for the incentive to go forward. I never use a kick because that is sometimes an abusive signal and sometimes results in a negative reaction from the horse.

Martha likes to lope a lot of circles in the pasture sometimes on the side of a hill. She squeezes the horse gently with her outside leg to urge him forward into the correct lead.

We do a lot of circles and sometimes do some reverse arcing at a lope. In a reverse arc, I lope a circle in one direction and hold the horse's head in the opposite direction. What I'm doing is taking the horse's head to the outside of the circle. If I'm loping a circle to the right, I would turn the horse's head to the left. Reverse arcing makes the horse's body supple and teaches the horse to allow the rider to put his body into several positions. This is a little added education that complements regular circles. It is also a position used to start a lead change.

We stay in the pasture until the horse can be completely controlled by me. There is really no definite time limit that any of these steps should take. Each horse is different. Once a horse has mastered one step, we move on to the next. For some horses, that may be a few days. For others, it may take weeks.

When I'm teaching a horse to take his proper leads, I prefer a big space because using cramped quarters is like using a crutch. I want to educate my horse so I can communicate in any situation, whether it be small or large. But for total communication, I have to put a horse in both situations for him to be a properly educated horse. He must be able to pick up the proper lead in a cramped space, a big space and in between.

Eventually I use single cues for lead changes, such as the outside flank or the inside foot or the inside rein. With education, I can get him to the point where by the end of his three-year-old year, he has accomplished picking up his correct lead from any one of these cues. The reason, I mention several different cues is that different situations call for different types of cues. For example, if you have to come around a left-hand gate to come into the arena, and have to pick up a right-hand lead to head for the first barrel, use a reverse arc to send your horse into that lead. If you're coming straight down an alley, it would just be the regular communication for the lead of your choice (whichever barrel you run to first).

When a horse is a finished barrel horse, I simply use the inside rein to put his head on the inside, and outside leg pressure to push him up into the proper lead. When he has the lead, to move his body into position to go around the right barrel, I use my inside leg. This needs to be reflex action on the horse's part as well as reflex action on the rider's part, for communication.

After you and your horse have progressed to running down a alley, you will find that each horse has a different time frame. But when he's working at a high rate of speed, it takes a horse three months to a year to retain the proper lead at all times. It's easy to put a horse into a lead, and then back in a lead after motion, if you've educated him properly; but for him to become seasoned to hold that lead, takes some time and experience.

If your horse is on the wrong lead coming down the alley, break back to a trot or crisply stop the horse, because the stop always makes you totally in control of the horse's body. When you totally control the body, you control the mind. I always want total control. After you go through the gate, most places have a rule that you can't stop forward motion. But if it happens when I'm approaching the alley, I change it right then. A horse cannot maneuver a turn as well in the incorrect lead and he cannot stay as sound if he runs in the incorrect lead. So he must learn to take the proper lead.

If you want to practice taking the proper lead, remember that practice has to be done perfectly. Practice that is sloppy doesn't help you, just as sloppy repetition doesn't help you. Perfect practice and perfect repetition are what help the horse and the jockey.

"THE HORSE WILL TELL YOU WHEN HE STARTS TO LEARN THE PATTERN."

7

Beginning Barrel Training

The band is playing and the rodeo announcer has just called your name. Your well-trained horse perks up his ears as though he knows he's about to perform for the stands, brimming over with an enthusiastic crowd. You race down the alley, headed toward three brightly painted barrels, skimming each barrel and leaving the third with a burst of power. The final trip home is accompanied by screams and cheering from the crowd. Your horse's muscles are rippling as he exerts his whole body, responding to your body language, which tells him to give it his all.

You cross the timer at a dead run and as you pull your horse up outside of the arena. You hear the announcer give your time in an excited voice. You don't know what everybody else's times were, but you can tell by the pitch of the announcer's voice, that you have run the best time. You lean over and give your horse a hug, get off and start walking him to cool him off.

Those are the dreams of every aspiring barrel racer. However, as in every sport, barrel racers must start with step number one and work their way up to the top. This can be done only if you know the proper way to either train a barrel horse yourself, or keep one that is already trained, solid.

Every two-year-old matures at a different rate, so there is no specific time in a colt's life that he should be started on barrels. It depends on the colt, whether he's mature and a quick learner, or immature, leggy and a slow learner. Slow learners should never see a barrel during their two-year-old year. However, we've had some colts that actually have gone around the barrels moderately fast as a two-year-old, not to beat the clock, but with a bit of run.

(overleaf) Horses learn by repetition, which in turn leads to habit. The Wrights introduce young horses to the barrel pattern by walking them to the barrel and taking them around the barrels the exact same way every time.

In the last chapter, you learned how to start your barrel prospect in the round pen and in the pasture. Once your colt totally handles what you want him to do in the pasture, with some speed and without you having to reinforce his memory every day, he is more than likely ready to start on barrels.

Each day when I ride a colt, I like him to go forward freely and softly, turn around, stop, back up and pick up his leads at whatever speed I want to do these maneuvers in — moderately fast to fast — simply off of cues that I have previously taught him and which we discussed in the previous chapter. I'll do some 360s (360-degree turns, spinning completely around) on him — that is 360s with forward motion, not 360s pivoting on the hindquarters like a reiner would do. I do 360s like running around a barrel. I want to be able to maneuver the horse in any position I want him to be in. If he can accomplish those maneuvers, then I feel he is ready for barrel training.

We've had colts that have never been on barrels and when people watch as we work them, they want to know how many months they've been on barrels. But the colts are working off cues that we taught them in Chapter 6. They were educated as to what we wanted them to do. With these cues, we can get the colt in the position we want him to be in before, and while, he is going around the barrel. He will look like a moderately green, barrel horse, just because he was educated before he got there.

Horses learn by repetition, which in turn forms a habit. We introduce young or green horses to the barrel pattern by walking or trotting them around the barrels. We take them around the barrels the exact same way each time, because changing any position does not form a habit. A young horse needs repetition.

Usually we trot between the barrels, then stop and walk around the barrels. Walking the entire pattern, even at first, is boring to a young horse and will make him lose interest if it is overdone. Walking, however, is very important to maintain a relaxed mind. But remember, even though you just walk and trot the barrel pattern, it is very important that you walk and trot the pattern correctly and the same every time.

If the colt walks around the barrels and does everything that I ask the first day that I go to the barrel pattern, then I just exercise him in the pasture and then put him up. We may not work the barrels again for three days. When we come back, I might push my luck. Walk the first time, trot the second time, and if he's done well, I'll quit. If he hasn't done well, I'll go back to a walk, reemphasizing what I want which is for a horse to move out when I ask him to, move over when I ask him to, and stop and turn when I ask him to, at whatever speed I ask him to do it. I always start by walking first and move up in speed. Start with the easy step before you reach the harder step.

If I have done my preliminary education right before we go to the barrels, the colt will work the barrels properly. If this is a situation where someone else was riding the

horse and didn't do their homework properly, I'll stay working the barrels, walking and trotting, until I get the horse to improve just a little. When he does, I reward him by quitting. If he doesn't improve, I have pushed him too fast or have not taken the proper steps along the way.

You can't take a horse off the track and then put him on barrels the next day. That colt won't do well because he doesn't have the foundation that we talked about in Chapter 6. He needs that foundation and education. If you have a colt that won't walk and trot around the barrels, you haven't provided the proper foundation. You need to back up and give him some foundation training.

It will probably be 10 days to three weeks before I slow lope a horse on barrels. I start by just loping between the barrels and slowing down to a walk or trot going around the barrels. Always, slow down when you get to a barrel, to reinforce the message of "rate" and "turn." The amount of time I spend on a colt each day is from 15 minutes (on a

a) When introducing a horse to the barrels, walk around the barrels and trot between them to keep the horse from becoming bored. As you head into the barrel, the horse should be about four feet away from it with his head tipped in toward the barrel.

b) Start your turn at point #1 in the pocket.

c) Maintain a four foot circle around the barrel and head for point #2.

d) At point #2 your horse has completed his turn. On the third barrel, you should over-turn the barrel so the horse does not become accustomed to leaving the arena before the turn is complete. In practice this barrel should be over-turned and you should head for the fence. See figure 7.1.

If your horse does not turn the barrel perfectly, take him around several times, but each time he goes around at a slow speed the pattern should be perfect. Ed trots slowly around the barrel, keeping an even distance from it, while tipping the horses nose into the barrel

Ed keeps a four foot pocket all the way around the barrel and continues to keep the horses nose tipped in toward the barrel.

Ed has gone all the way around the barrel, but circles again to make sure that his young horse learns to turn the barrel correctly through repetition.

highly receptive horse), to an hour. But I never spend that much time on the barrels. You never want to have to tire a horse out to teach him something. That teaches the horse to focus on you only when he is fatigued.

If I have to fatigue a horse to make him work correctly, I haven't educated the horse properly. Fatigue is a form of intimidation training. It can be used to discipline, but if it is used every time, it is intimidation training. My only goal each day is to get the horse to walk and trot around the barrels, with me controlling every move and reaction of the horse and doing it the same way every day. That forms a perfect habit for the horse.

TEACHING RATE

Rating is when a horse collects and shortens his stride. If he's trotting before you reach the barrel, he'll slow down to a walk. If he's loping to the barrel, he'll automatically slow down to a trot all by himself. If you are just walking to a barrel, the horse should be willing to stop, then go forward and go around the barrel at the same speed.

Many people get in trouble with their horses when they walk the pattern and the horse does it well. So then they trot the pattern, and they are still doing the pattern in perfect position. They might even get by with slow loping and have a perfect barrel. However, when they move up in speed, and they haven't taught a horse to rate and stop, they'll overrun the barrel and teach the horse a bad habit. One cardinal rule to remember is that whatever speed you go to a barrel, you need to slow down to the notch below that to go around the barrel. That teaches a horse rate.

To begin to teach your horse rate, trot to the barrel, then stop and back your horse up about 12 to 15 feet before the barrel. This is the spot that you would ordinarily "rate" or slow down your horse if you were loping him, so start the repetition now. You do this until a horse starts to rate, and keeps rating, on his own. Then there's no need to stop each time. The place where you should rate your horse is the spot where your horse's tracks crisscross each other (coming in and going away from the barrel). However, when you are starting a young horse, your rate position should be sooner than that (see figure 7.1 on next page).

THE POCKET

Just as you should practice your rate when you start a horse on barrels, you should also practice making your pocket when going around the barrel. The pocket is the distance from the side of the horse or rider's stirrup to the barrel. The pocket is taught to the horse so he does not touch the barrel while maintaining forward motion. The pocket should be up to four feet in depth when you first start working the pattern. That same day, after you have gone around the barrels several times, you can decrease the pocket size to two feet in depth.

If we come to the barrel in preparation of making a regular pocket, and the horse wants to turn so much that

Martha reaches the rate position on the second barrel. Even if you are walking a horse on the barrel pattern, he should be rated at the same place every time. After a while, the horse will automatically slow down to a trot or walk by himself.

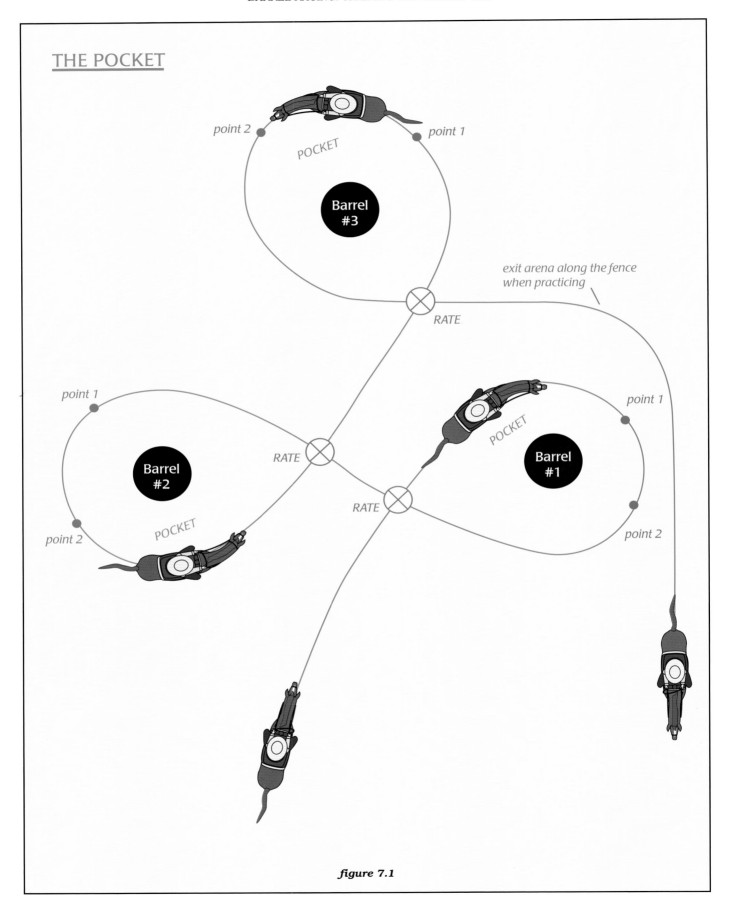

figure 7.1

he closes the pocket down too small, we increase the size of the pocket to from six to 12 feet, so the horse will honor the two to four-foot pocket. Then, when the horse starts honoring the six to 12-foot pocket, go back to the four-foot pocket. If he closes in again, move back out and repeat this procedure until the horse honors the small pocket.

The horse will tell you when he starts to learn the pattern. When he starts to make some moves correctly on his own, then let him do five percent more on his own. If you're headed to the barrel and you feel the horse rate himself, don't rate him. Don't enhance it. If you feel him starting to turn, be in a position to help him. If he wants to do it, just help a little — not totally. Each day that he learns more, do less for him. The more the horse does, the less the jockey should do. However, this happens in small degrees.

As you approach and turn each barrel, use the reference points (in figure 7.1) on the ground at the spots where you begin and complete your turn. This will simplify the steps taken in making each turn. Point #1 is the reference point on the ground where you should start your turn. You should then progress forward to point #2, where you actually complete your turn. Point #2 is the spot on the ground where you ask your horse to complete or finish the turn and drive away from the barrel.

After your horse's body has completed the turn, head for the next barrel. Repeat the same exercise on each barrel. Trot to the barrel, stop, back up a couple of steps; then go forward at a walk, making your pocket, round the barrel, then head for the next barrel at a trot. Make sure your horse has completely turned the barrel. Remember, as the horse starts rating on his own, quit stopping him because you can get too much rate on a horse.

WHERE TO LOOK

Even at this slow pace, where you look is important. As you leave the alleyway, look at the spot where you will rate, or stop, your horse. If you look at the barrel, you will tend to drift toward it and you'll more than likely hit it or make too large of a pocket if something out of the ordinary occurs. Again, as you head toward the barrel, when you stop or slow down your horse to rate him, look at point #1. When you have started your pocket, look around the barrel at point #2 and after leaving point #2, you should be looking at where you will rate or slow down for the next barrel.

Each day, take into consideration at what point the horse was the last time you had him on barrels. If he was trotting the barrels and doing a lot of maneuvers on his own, (which we call "putting him on his honor"), then move up to a slow lope and slow him to a trot at the barrel. If he picks up the turn on his own, he has advanced.

WHEN IS IT TIME FOR ADVANCEMENT?

Certain horses can take more advancement in one day, but we don't suggest this because not many riders know what is "too much." Too much is highly destructive, where

a) Ed reaches the rate position for the second barrel.
b) Ed creates a pocket by moving his horse to the right with pressure from his left leg.
c) Ed has tipped his horse's head in toward the barrel and reached point #1 where he is starting his turn. His left leg is still pressed against the horse to maintain his pocket.
(opposite) figure 7.1 Your horse should leave a four foot pocket around each barrel. Rate him where the paths cross. Point #1 indicates where the turn starts and point #2 indicates where the turn is completed.

The horse is out of position between the first and second barrel. When heading toward the second barrel, the horse should not be to the left of the barrels.

not enough just doesn't advance the horse as fast. You won't destroy what you have already taught him. The average horse should start rating on his own, at a walk or a trot, within a few days, or maybe even the first day. However, it can take from 60-90 days to rate and turn on his own with 50 percent of his full speed.

When a horse does everything you want him to do, such as rate and turn at a very controlled speed, then it's time to move him up a notch in speed and get a little quicker in the turn. This can be done by urging him forward with the calves of your legs, by your body position being forward or even tapping him on the hind end with something. But anytime you step up in speed, it's also time to back off and take the pressure off your horse. If you have worked one time at a faster rate of speed, the next time you work the horse (the same day), back off on the speed. It's easy to put "speed" in the mind of a horse and hard to keep "slow" in his mind. Always quit on a slow work.

If you move up a notch in speed and the horse doesn't allow you to control him, you are moving too fast. Or, if you're "putting a horse on his honor" and he doesn't retain the rate and turn that you have taught him, you have gone too fast. If he takes one step out of position, you need to slow back down and put it back in his mind right then. Teach him that he needs to do it right.

Don't quit with a problem. You must quit with a positive work. You can't get in trouble by staying slow too long, but you can get in lots of trouble by progressing too fast. Riding in the pasture for a few days, away from the barrels, or turning the horse out periodically allows his mind to stay fresh. You want to keep it fun for your horse by not advancing too fast.

Some horses can take fast advances and a lot of repetition. Some can only take a little bit each day; otherwise, it becomes boring or stressful. Every horse is a little different. When your horse is bored, you'll more than likely feel it. He won't want to excel. He won't try for you or give you his attention. Also, when he's bored, he'll start looking off, feeling a little unresponsive underneath you, or he may even get a little tense.

Be alert to the subtle things the horse is telling you. A stressful horse will let you know he's stressed. He'll get a little heavy in the face by pulling on the bit, grind his teeth, stomp his feet or even pop his nose or tail. These are the most conspicuous signs. But there are also some less obvious signs.

A horse is stressed when he wants to go forward at a speed faster than you want him to go. He's also stressed when he doesn't hold the turn. This means that he takes one or more steps away from the turn, out of his perfect pocket. If he turns before he gets to the barrel or rates before he should, those are also signs of stress.

Other signs of stress are overreacting to the turn (by turning before he gets to the barrel), or turning around at the beginning of the pattern and not wanting to go back to the pattern. Also, when he has turned the third barrel, and then acts like he doesn't want to return to the starting position, you could have a stressed horse on your hands and you need to back off and go slower.

All of these stressful signs mean that your horse is dreading the pattern. You haven't made the barrel pattern positive for him. You have either overdone the pattern or done something to make a negative impression on him. He obviously doesn't understand what you're trying to get him to do and, therefore, it's time to back off and quit.

There are so many of these signs to watch for. Besides the signs we previously mentioned, you can also tell by a horse's ears if he's stressed. If he flicks his ears back and forth like something is biting him or if he just pins his ears (lays them back against his head). Also, if you have to urge him forward and urge him to run because he seems to be resistant, it could be caused by stress.

We do not believe in asking for speed too early. Speed can be your enemy. Precision, repetition and letting the horse do it on his own (putting him on his honor) are better. We believe in putting a horse in a position so he can do the barrel pattern as much on his own as possible. But each horse is different. Some horses will always have to be helped, while others reach a mature stage early when you don't have to help them much at all. The latter is surely the better horse to train and eventually sell or win on. This will be discussed further in the Chapter 8.

When your colt works the barrels softly and easily with no stress or resistance, and then rides away from the training session relaxed and easy, you should quit for the day. For most colts, it's hard to get them totally focused with only one work. They need repetition, just as people do. But it's important that your training session is productive, pleasant and as fun for your horse as it is for you.

CONTROLLING THE HORSE

You can control a horse with eight parts of your body: two hands, two heels, the calves of your two legs, your rear end and your body throttle. You can control the horse with your hands by using them as cues. Take the inside rein and push it against the horse's neck in the opposite direction. This moves him out and bends his body.

You can pick up the reins with your hand to slow him down and you can lift the reins with your hand in the air to pick up the horse's shoulder. You can also lower your hand to put the front end of the horse a little lower, or you can use your outside hand to keep him from pushing his shoulder or rear end out. You can use your hands to slow a horse down by softly touching or bumping the reins, one side of the mouth at a time. If you pull on the reins with both hands equally, you make a horse want to lean into, or push against, the bit.

Never take a straight, solid hold of a horse's face at any time, regardless of whether or not he is wearing a bit or a hackamore. Seesawing the reins back and forth, from hand to hand, keeps a horse soft in the mouth and the body. It's better to work on one side of the mouth at a time. That way, the leg comes under the horse on the side you are working. Then, if you work on the other side of the mouth, the other leg comes under. Eventually, the horse will become educated enough to put both legs under him at the proper time to stop.

Where your hand is means something to a horse. On a horse with an elevated head, or a higher head set, lower your hand position and stop him with a give-and-take action with your wrist and little finger. That method is fast and has feeling. Using the whole arm is slow, heavy and strong. On a horse that wants to be a little too low-headed, elevate your hand position slightly.

You can also use your hands on the reins to balance a horse. The outside rein is called the "balance" rein, because if it is picked up lightly, it can keep balance in the horse. The inside rein is the "control" rein. The balance of the outside rein is a very light touch to keep the horse's backbone lined up in position. The outside rein should be

(opposite) Good and bad hand positions.
a) The rider's hands should be spread further apart.
b) Good hand position in the turn.
c) Good hand position for picking a horse's shoulder and putting the horse's rear end under him.
d) A lower hand position will bring the horse's head down a little. Use this position if the horse is extremely collected.
e) The reins are too long and the rider has let her hand slide too far back on the reins. Also, the hand is too high.
f) A good hand position at point #2 on the barrel pattern.
g) Good hand position at point #1 on the barrel pattern.

BEGINNING BARREL TRAINING

91

directly above the outside shoulder, with very light contact with the mouth. It isn't a pull; it's a light contact. If you use that outside rein, the horse doesn't move his shoulder or rear end out or go forward with his head up and back. When you get a horse fully educated, hopefully, you can control him totally with the inside rein. However, the outside, or balance rein, is necessary in educating the young horse, as well as being necessary on a lot of horses used later in competition.

Your legs also give cues to the horse. For forward motion, ride the horse forward with pressure from both legs. Use pressure from the inside leg on the horse's rib cage to get the horse to move and bend his body to the outside. Use pressure from the outside leg on the back of the rib cage to keep him from swinging his rear end.

Your legs are also used to turn a barrel. Lay your inside leg (the leg closest to the barrel) on the horse's rib to build your pocket, flex your horse and get supple in the rib and shoulder. (Supple means to move the shoulder and rib out three to four inches in a position to go around the barrel with a slight arc or flex in the body.)

Your heels are used to propel the horse forward. We almost always wear spurs, but we use them very lightly. We never spur a horse for intimidation; we only spur as a line of communication — to move over, either away or toward a barrel, or both used at the same time for forward mobility.

Your rear end in the saddle is attached to the throttle. The feel of your rear end in the middle of the saddle means slow down to the horse, while sitting toward the back of the saddle means "whoa." "Body throttle" is like body English. Putting your weight on the front part of your saddle is the cue for a horse to go forward. Sitting upright and putting your weight in the middle of the saddle is a cue for the horse to set or come back to you (slow down a little).

When you sit up in the middle of your horse, you can feel what he's about to do and go with him. If you're weight is back too far toward the cantle, you'll only feel what the horse has already done. You'll be behind his motion instead

a) Martha trots the barrel pattern, using her inside "control" rein to pick up the horse's inside shoulder and keep him from getting too close to the barrel.

b) Martha keeps her inside rein to maintain her distance from the barrel.

c) Martha uses a little of her outside "balance" rein to keep the horse's backbone lined up in position. Once she has finished her turn, she will be in position to head straight for the next barrel.

of keeping in time with it. In a speed event, you don't need to be totally in the back of the saddle unless a horse is really green. Then you move back farther to the cantle. That's the only time you should ever be against the cantle.

When you know what your horse is going to do before he does it, or when he feels what you want him to do before you ask him, that's when you're "with a horse." You and your horse are one. Barrel racing is a one-unit event and that unit is a combination of the horse and rider. You should not be separate individuals. You should work as a single unit — a team.

Therefore, you use your hands, legs, rear-end and body throttle according to what the horse is doing. If he's setting (or rating) too much, move forward a little and urge him with your legs. If he's drifting out, use the calf of your outside leg to push him in. If he doesn't respond, tap him with your heel. The outside hand on the rein can also be used to balance the horse. This means that if you are pulling a horse around to the right, also keep some contact on the left rein to balance the horse.

If you are an experienced rider, it doesn't hurt a thing to help a horse, but use finesse. We're not talking about running up to the barrel, pulling on the reins and jamming the horse into the ground. Ride with "hands on," meaning you have contact with and control of the horse. We don't let a student's horse be "on his honor" much because we want the student to learn to put his horse in the position he needs to be in. A student needs to educate his horse to get in that position on his own. The goal is to produce an automatic horse. However, horses that never achieve that peak can still be competitive through your education of the rider.

THE TURN

There are three different styles of turning barrels. Each horse is different and has his own individual style. The style that a horse is comfortable in turning is what the rider must work with. But regardless of which style your

horse uses to turn a barrel, he must still set (slow down and gather himself) in order to utilize his feet around the turn and take the proper lead during the first step away from it. If he sets too much, he either loses too much time or he hits a barrel.

The Four-Wheel Drive Turn: We prefer this particular style of turning because we feel it is quicker, easier to control, and much easier on the horse, both physically and mentally, as well as the rider. With the four-wheel drive turn, a horse runs to the barrel, collects and rates, (slows down) just enough to totally control the turn and never steps out of the pocket on any part of the turn. He uses all four legs equally, pulling with both front ones and pushing with both back ones separately, and he's spread out for balance. With this style of turn, the horse has equal weight (or usage) on all four feet and his body seems to drop down close to the ground as he runs around the barrel.

This is what we call a "flat-turning" horse. The horse does not lean into the barrel nor arc his body away from the barrel. He simply flexes his body around the barrel and runs around it, using all four feet for propulsion. The backbone of the horse appears to be slightly bent while he is turning the barrel.

In a four-wheel drive turn the horse uses all four legs equally and drops his body down close to the ground as he rounds the barrel.

In a rear-end turn, half to three-quarters of the way around the barrel, the horse sets, pauses, rolls back over the inside back leg and pushes off to the next barrel.

Not enough flex and too much flex in a turn are both counterproductive. We like some flex in the turn; however, if we had a choice between a horse that was as stiff as a board and one that is bent like a noodle, we'd take the one that was stiff as a board. When a horse bends like a noodle, he loses his ability to go forward and can injure himself easily by bending excessively.

Our goal is to have a horse that is balanced. Charmayne James' great horse Scamper, that carried her to 10 world championships is extremely balanced in competition. He runs in toward the barrel, stays on all four feet, moderately flat, and moves his body slightly into position as he turns around the barrel. He rarely has ground trouble because he stands straight up on his feet in a balanced position. He works, even if he loses his bridle coming in the gate like he did in the 1985 National Finals Rodeo. He turned all three barrels in perfect form, winning the go-round and the world championship.

But it's hard for most horses to simulate that turn. They really have to be athletes. Some people say that Scamper doesn't rate, and it doesn't look like it when you watch him run. But if you watch him run on a slow-motion video camera, you see dirt fly up as he approaches each barrel. Charmayne doesn't give him an obvious cue; he rates himself, which is what we'd like all horses to do.

The Rear-End Turn: This is the next most advantageous turn. In the rear-end turn, a horse sets at the proper place prior to turning the barrel. His backbone is fairly straight, without much flex or bend. He goes from one-half to three-quarters of the way around the barrel, pauses, and rolls back over the inside back leg, pushes off and goes to the next barrel.

However, there are two things wrong with a rear-end turn. The horse has to be a superior athlete to win, as there is no adjustment for error. You've got to go into the barrel in perfect position because you can't move your horse in or out very much. (In the rear-end turn, you don't make a pocket where you can move your horse closer of farther away from the barrel.) Also, it's really physically hard on a horse to lock down on his rear end, stop, turn around and drive off again.

Front-End Turn: This is the least desirable of the three turns. In this turn, a horse runs up to the barrel, keeping his backbone extremely straight and stiff. He then runs three-quarters of the way around the barrel, drops the front end, moves his butt over, but only until it is in alignment to go straight to the pocket for the second barrel, and then runs to the next barrel. A horse that turns on his front end and runs around the barrel is really hard to position and balance, because — just like the rear-end turn — you can't move him in or out from the barrel when you're at a dead run.

But there are some highly successful barrel horses that naturally turn on their front ends. We knew a barrel racer whose horse would run up to the barrels, throw his head down, stick his front end in the ground, swing his rear end and take off again. He carried his rider to a world championship. The first time we saw him run, we thought the clock didn't work. The second time we saw him run, we thought the clock was surely messed up; that he must have a magnetic field around him. But after watching him more, we figured out that he just out ran the competition. The horse was tough, but he had a freakish, front-end style of turning.

That just goes to show that if you find a horse that has a style all his own, don't try to change him to what the perfect style is, because that is the perfect style for him. You may never find a barrel horse like Scamper, but you can still win on a horse that you have helped with proper training. Even though there are only three main styles of turning, in reality there could be numerous different turns, if you take a part of one of these turns and mix it with parts of another. The three we mentioned are the only "pure" styles of turning.

A horse's style of turning is natural, but you can determine early in your horse's training which style he uses when you are educating him to turn the barrels and then enhance and perfect it. When you determine the horse's mental and physical capabilities in the round pen and in the pasture, you may be able to determine which kind of style he will have. He may reveal that style when he lopes

When a horse turns on his front end, he keeps his backbone extremely straight and stiff, drops his front end and disengages is rear end to some degree.

circles, stops and turns around, and then you will educate him along those lines. You can change a horse from his natural style to another style, but he won't be as tough as he would be if he were left to turn his natural way. So if you have discovered what the natural style is for your horse, simply help him to turn naturally in that style, rather than try to change him.

CORRECTING THE HORSE

When correcting a horse that has done something wrong, try to stay out of his face. Don't jerk or pull on the reins. There's a half-inch difference in the way we "handle" a horse and the way we "release" a horse. "Handle" means the light touch of the hands on the reins to communicate with the horse. "Release" means not using the reins to communicate and releasing pressure on the reins, depending on how well the horse is educated for a perfect turn and run. You need to use finesse, which means asking, not forcing, a horse to do something. Handling a horse is giving and taking, through the reins, on the corners of the horse's mouth. Your reins are your telephone lines to the horse.

When I touch the reins softly with my hands and don't get a reaction off of that taut rein, I'll take eight or 10 inches more rein, which puts pressure on the mouth of the horse, until he gets light and yields to me. When he does, I release the reins immediately. Then I put another soft touch on the reins. If he doesn't react, I take another 10 inches, putting pressure on the mouth, until the horse gets light and yields. Then I release the reins again. That teaches a horse to get soft in the mouth, because as soon as he gets soft, I will give him the reward of releasing all of the pressure. I never scold a horse with a loose rein.

To correct a horse with my foot, I'll touch the horse with my heel. If I don't get a reaction, I'll roll the spur up his side eight or 10 inches, briskly and aggressively. I never jab him hard, I just put my spur against him and either hold it there or roll it up his side. I always use a spur to educate a horse to maneuvers. If a horse doesn't have light sides, where only a heel (without a spur) keeps him soft (and 95 percent of the horses don't), I use a spur. But roll it up a horse's side, never jab him with it.

STRIDE

The "stride" of a horse is a full cycle of limb motion — the distance covered by one foot when in motion. If the distance is short, he is a short-strided horse. If the distance is longer, he is a long-strided horse. You can't change a horse's stride from the way it is naturally. Great horses stride short and long when they need to. For those that are choppy, you can only slightly increase the length of their stride. For those that are long-strided, you can shorten the stride slightly. But you always have to be educating a horse that is not naturally strided the way you want him to be. The unnatural way must always be enhanced in their mind.

A horse needs to be short strided so he can round the barrels efficiently, and long strided so he can make up time between the barrels. Ed teaches his colts to lenghten their stride by long trotting them.

This colt has a long stride, but also needs to learn how to move with a short stride. Martha trots and canters him slowly to accomplish this.

Stride isn't something a horse just learns and then you can quit working on it. It's a continuous process and has to be re-enhanced often. If your horse naturally has an extremely long stride, he usually won't mentally pick up on being short-strided. Only a low percentage of long-strided horses will pick up on being short-strided around a turn, where they need to be. To lengthen the stride on a horse that is naturally short-strided, between barrels, you usually have to keep re-enhancing it in his mind also.

We long trot a short-strided horse to help lengthen his stride, as much as his body or conformation type allows. In a long trot, a horse covers a lot of ground, as opposed to a slow, jog-trot, in which the horse barely makes any headway, but does move in a slow, comfortable fashion. A jog is often called a "sitting" trot, simply because it's easy for the rider to sit. In a long trot, the horse lengthens his gait by extending his body as he covers ground. The pace is usually fairly fast and it requires the rider to post or stand up in the stirrups to handle the motion of the horse.

With a horse which is naturally long-strided, we try to shorten the length of his stride. This is done by trotting him slowly and cantering him slowly and in a collected fashion. A barrel horse needs to have a long stride between the barrels and a short stride around the barrels, so an ideal barrel horse should naturally have both.

When we exercise a horse in the pasture, we like to vary the length of his stride. In our conditioning program, we might start out by working the horse first in a long trot and then asking for a canter, or slow lope. We condition in this manner for a half mile and then work up to five or six miles every other day, with short work in between. Finally, we might even sprint (run at a fast rate of speed) for a short distance, approximately 50 or 100 yards.

We also exercise our horses for variation in the length of the stride on the barrel pattern. We do this by loping to the barrel, short trot (or jog) around it (not long trot) and then lope to the next barrel. We always shorten the stride going around the barrel. When we talk about "collection," in going around the barrel, we are talking about shortening the horse's stride, not slowing the horse down. You have to shorten the horse's stride to go around the barrel and leave it the quickest and hardest way possible. A long stride is not feasible around the barrel because it automatically moves the horse out from the turn.

When we purchase a horse, we really pay attention to his stride. To check the horse's stride, put a horse in the round pen and turn him on the fence. See if he collects himself by shortening his stride or if he tries to reach with his front and back feet as he is turning and leaving. We want one that collects his body when he turns. When you send him around the round pen in motion, see if he lengthens his stride like he should. Also, when you are riding him outside in a circle, see if he has the ability to shorten his stride for a turn or if he has to reach out all of the time. We want a horse that has the ability to shorten his stride.

It is hard to state a time frame for beginning barrel horse training. You train a horse using his mental capacity to learn and retain in his mind. He can learn only as fast as he can retain. Even though he fulfills what you want him to do that day, you need to make sure he retains it during the next session. Consequently, you advance as the horse accepts and retains mental input.

The process mentioned in this chapter can take from 40 days to 10 months, depending on the horse. We had a horse that, within 40 days after we got her from the race track, could win a barrel futurity. It wasn't because she was a finished barrel horse. She was smart and listened to her rider who was good enough to train her. With other horses, it has taken 10 months before we could run a barrel pattern on them.

Each horse has a different time frame for advancement. Also, the person training the horse dictates the speed at which a horse learns. Many people are not capable of training a horse. In fact, if they are not educated themselves on how to train a barrel horse, they may actually do harm to the horse's training program and set him back. If this happens, when he finally gets in the hands of an experienced trainer, that trainer will usually have to start from the beginning and put a foundation on the horse.

"RUNNING FAST IS THE MOST NATURAL INSTINCT THAT A HORSE HAS."

8

Increasing Your Speed

When teaching a young horse the barrel pattern, the work should be done at a walk, trot or lope. After the horse is mentally and physically advanced enough to handle the pressures of moving at greater speed, it doesn't scare him and he remains under control, then you can let him run between the barrels with a small increase in speed. Confidence with speed is developed in the barrel pattern through progression of speed. Progression means trotting the pattern until the horse understands, and remembers through repetition, the perfect position that the rider has repeatedly shown the horse. Repetition is the key!

Then move up to the slow lope, fast lope, slow run and fast run. The next step is taken only after the horse understands and remembers the current step. Keep in mind that if you move up in your training program from a slow lope to a fast lope, and the horse doesn't do his pattern exactly as you want him to, take him back to the slow lope. Never move up a notch in speed unless the horse is doing the pattern perfectly by running at the previous speed. Also, at some point in the speed scale, the speed may "blow the horse's mind," meaning that he can no longer do a perfect pattern with any speed. If this happens, you need to go back to the basics and start over. In fact, you probably need to go back to riding in the pasture for a while.

WHEN DO YOU LET A HORSE RUN?

Horses that have never been to the race track usually have to learn how to run. The learning must be done in such a way that the horse doesn't become uncontrollable in the process. You move him up to speed slowly. The progression is done from long trotting and loping to galloping at a maximum speed, but with control. The final

(overleaf) Most horses don't naturally know how to run properly and have to learn how. The secret to training a barrel horse is being able to control that run.

Sprinting a horse will free him up, mentally and physically.

step is teaching a horse to sprint (running 20-50 yards as fast as possible). This is usually used to mentally "free up" a laid-back horse. It puts the "fun" into "run." But when you teach a horse to sprint, make sure he maintains the proper lead and that he reins and stops well. In other words, make sure he is thinking, and remembering his training.

Running fast is the most natural instinct that a horse has, especially those bred "hot" enough to be barrel horses. They know how to run and love to run. The secret is controlling that run. Teaching a horse to run doesn't mean working him up to running as fast as he can go and then letting him run at that speed from then on. Teaching a horse to run means controlling the horse to run at the speed you want him to run, while responding to your cues. As long as your horse is working properly, responding to your cues, and doing what you ask him, you can let the horse advance in speed. But anytime the response from the horse is not what you are looking for, you need to back off on the speed.

When you sprint a horse for the first time, he should be in good condition physically and mentally. Physically, the horse should be prepared by an exercise and conditioning program. You should have been loping him every day and his lungs and muscles should be able to take a run. Just like you wouldn't race a mile yourself when you weren't in top physical shape, you shouldn't expect a horse to either. If you do, you could easily cripple the horse by tearing a muscle or a ligament.

The horse should be prepared mentally through education and training, so he responds to your requests without getting scared and nervous. You should be able to rein and stop him easily before any speed is incorporated into his program. We don't suggest sprinting on a regular basis, just enough so that the horse realizes that he will be allowed to run as fast as he can run for a short distance. The ground should be safe, not too hard or too soft. It should have some traction and serve as a shock absorber for the horse.

The sprint should only be 20 to 50 yards, just a short distance, like you would run in a barrel race. However, some really laid-back horses need more distance to "free them up mentally" (make them willing to run and do so easily), when requested to. Also, at the end of the sprint, don't stop the horse suddenly until you have slowed him down to 20 to 30 percent of the speed he was traveling. A sudden stop at fast speeds could hurt the horse's mouth through severe rein pressure, or his legs due to a bad stop in hard ground conditions.

In the beginning, when you are teaching your horse to run, let him pick his own speed. Don't urge him to go faster or slower until he learns how and when to use speed. When he starts to respond, you might ask him for a fraction more, by urging him with your legs, clucking to him or even patting him on the butt with a quirt, so he understands what you want and accepts the pace. Once you have taught the horse to run, and how to control the speed, you are ready to use that speed in the barrel pattern.

THE FIRST BARREL

Using speed while going to the first barrel can present a problem if you're not an educated rider with a solid horse. This is because you are coming toward it at the highest rate of speed of any of the barrels that you approach. However, the first barrel is not just a problem because you have more momentum. Sometimes you are headed to the barrel from the wrong angle. Therefore, prior to your run, you should study the angles to the first barrel.

Where you look is where you go. Therefore, do not look at the first barrel when you are running to it. Look at the spot on the ground just before the barrel. This is called the "approach" area. Don't ignore the barrel. In your "mind's eye," you know where the barrel is, but look at the approach area. Also, in your mind's eye, you know exactly where your horse's front feet are. They're right below your saddle horn, so use your horse's front feet as the marker for the approach area. The horse's front feet are the rider's location point on the horse.

Also, don't run straight at the barrel — run in a slight arc. If you run to the barrel in a straight line, you would have to move over to get in the proper position and make an obviously harder, tighter turn. We don't like hard turns; we like smooth, easy, symmetrical turns with speed involved in them. Any time you make a hard turn, you have to slow down. A hard turn requires more of an advanced stop. A smooth transition allows forward motion while you are turning the barrel.

RATING THE HORSE

Since rate is the degree to which you slow down or control the forward motion of the horse before you turn the barrel, it can vary from horse to horse. However, every barrel horse must have "rate" to turn the barrel. Rating can be compared to turning a corner in a car. You must slow down enough to turn the corner, but still keep up enough momentum to get it done. How to teach a horse to rate was fully discussed in Chapter 7; however, with increased speed, this is one part of barrel horse training that needs to be reinforced by "going back to the basics" until the horse does it automatically on his own every time.

Since some horses never rate on their own every time you run barrels, this is a maneuver that the rider must help the horse with, sometimes throughout their barrel racing career. When you come through the arena gate on your horse, you've got to feel how your horse is. If he's "chargey," you've got to rate him a little more. If he's lethargic and laid back, you've got to send him or push him a little harder. Rating a horse requires concentration and decision on your part.

Where you rate is called the "vital approach" area. This means that, regardless of what speed you are running the barrels, you must be in perfect position in this area and hold your rate until you are totally around the barrel and facing the approach area of the next barrel. In addition to rating your horse physically, it's also necessary to rate your

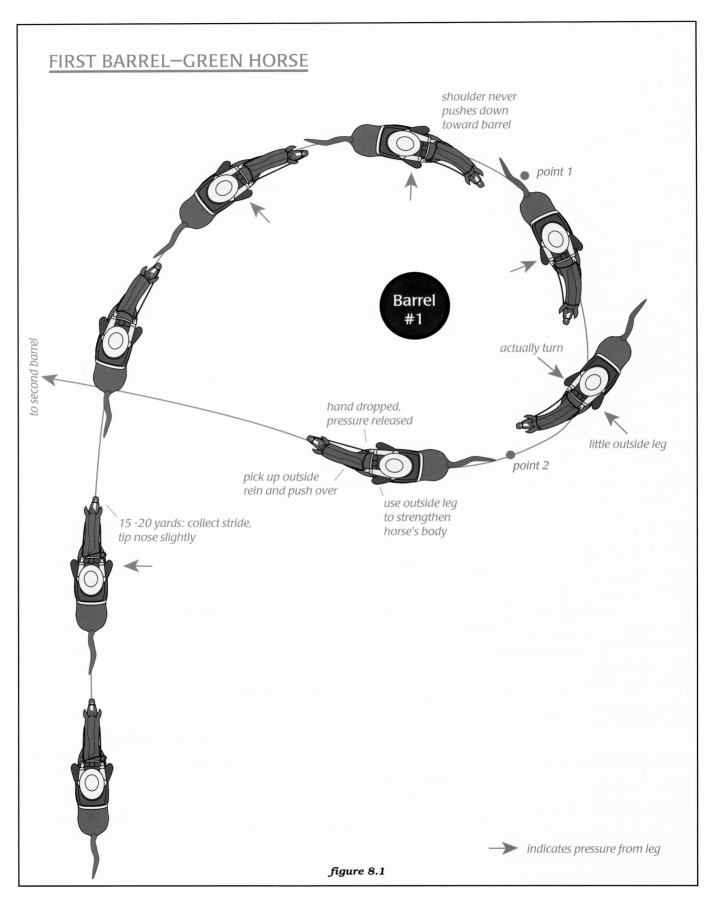

figure 8.1

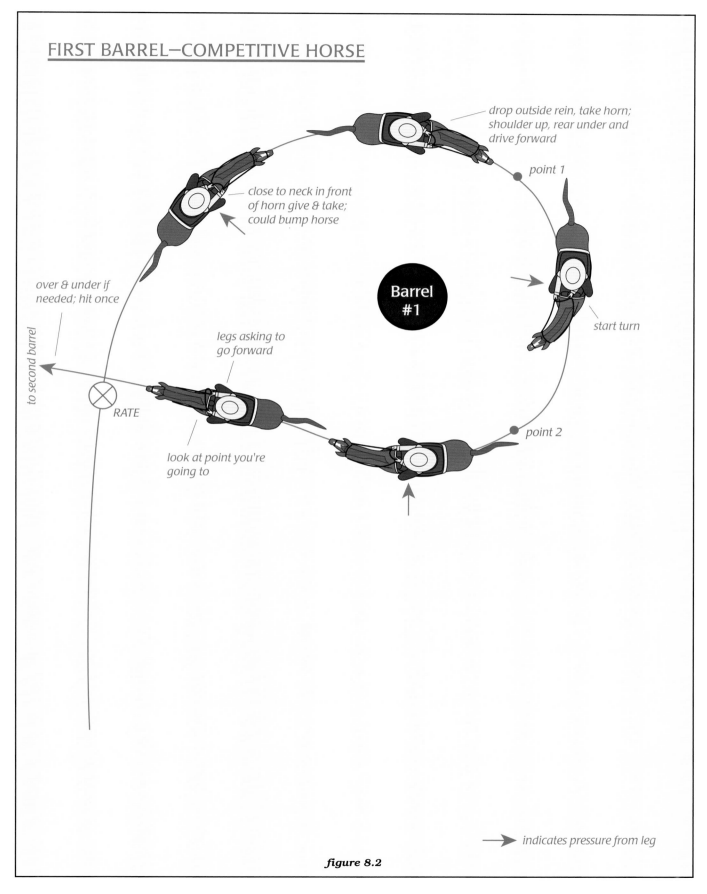
figure 8.2

(page 104) figure 8.1 On a green horse you turn the first barrel riding two-handed and and using pressure from your inside leg all the way around the barrel, until you reach point #2 where the turn is complete. Then your outside leg should exert pressure on the horse to line him up straight for the second barrel. You have to physically rate your horse 15-20 yards prior to the first barrel and tip his nose slightly inward.

(page 105) figure 8.2 On a competitive horse you hold on to the horn with the left hand as you race toward the first barrel. Your right hand holds the reins close to the horse's neck in front of the horn. Once around the barrel, use both legs to ask the horse to race forward, and look toward the spot where the horse will start his turn into the next barrel.

The inside rein is the primary rein and should be slightly tighter than the outside rein. The loose outside rein shown here is used for balance if the horse over bends.

horse mentally. You do this by consistently giving your horse the cue to rate. Repetition helps him form a habit; therefore, be sure you rate your horse in the same place every time with a clear and obvious cue, regardless of whether you are going fast or slow. If you completely stop your horse and then go forward, sometimes that actually teaches a horse to "go through your hands," or keep going. But if you stop him and back him up a step or two, it holds "set" in his mind.

If a horse wants to run by the first barrel when you are practicing, rate him early. Once you rate him, keep him slow all the way to the barrel. Then the next time you take the horse to that spot, and he rates, move the "rate point" closer to the barrel. Of course, when you're in competition, you can't rate a horse 10-15 feet from the barrel and still be competitive. Therefore, after you have the horse educated to rate, and you feel him beginning to rate on his own, move the "point of rate" closer to the barrel. Your goal is for the horse to eventually rate at the "win" position, which is quite close to the barrel. If a horse really throttles down before the barrel, he just can't compete with the horse that runs all the way to the barrel and then rates.

Regardless of what speed you are going to the barrel, rate your horse by utilizing both reins to achieve the proper position desired for the horse to rate and go around the barrel. The inside rein is your primary rein while the outside rein is your "balance" rein. Your primary, or inside, rein can be slightly tighter than the outside rein, just enough to slightly cock the horse's head to the inside. If the horse over-bends, you can "balance" him by gently pulling on the outside rein.

Once you arrive at your "rate point," and have rated your horse, you release your outside rein and rein the horse around the barrel with one rein only - the inside rein. We don't like to rate a horse totally by just using one hand because that usually pulls a horse out of position. You'll either be pulling his head in or his head out. The only time you would rate a horse with one hand is if the horse is a finished (perfectly educated), solid barrel horse.

Once your horse rates, as you approach the barrel, you want your horse slightly flexed in the area of the shoulder and upper rib cage or what we call "in position to turn." You are actually half or three-quarters of the way around the barrel (point #2), before you actually ask your horse to turn the barrel. Never look at the barrel because you will have a tendency to go where you look which means you would probably hit the barrel. Look at the ground in front of you, not right under your horse. Look at point #1 when you rate and when you reach point #1, you should be looking around the barrel for point #2. After leaving point #2, look up at the approach area for the second barrel.

The stride of a horse plays a part in where you rate. A long-strided horse, running fast, should be rated more firmly and a fraction sooner, than a quick, short-strided horse because it takes him longer to gather himself up to rate and go around the barrel.

Some top barrel racers make rating their horses look so easy because they have broke horses. It looks like they just get up there and ask their horses to run. An amateur or beginning barrel racer might think that the best horse runs clear to the back side of the barrel before he turns. But that's not always the case. Many of the top horses rate themselves. They're on "automatic pilot."

THE POCKET

The size of the pocket depends on the size of the horse and his athletic ability as well as the speed you are traveling going around the barrel. However, the size of the horse is not always a factor because not all big horses are long, tall and clumsy and not all little horses are quick. The ability to go around the barrel as quickly as possible should dictate the size of pocket you use on your barrel pattern. The size of the pocket should use up as little room as possible and can vary from two and one-half feet to six feet. The slower your speed around the barrel, the larger your pocket. When you take a barrel turn fast, the pocket becomes smaller, as the momentum of the horse running moves him in.

But you must strive for a pocket that allows your horse to travel forward the best. If a small pocket shuts your horse down too much and causes him to lose too much time, give him a larger pocket so he can keep his forward motion. There shouldn't be a lot of hesitation in the turn. The correct way to turn a barrel at any speed is not stopping and turning; it's running around the barrel and going on to the next one.

LEAVING THE BARREL

After you turn the first barrel, look ahead at the spot on the ground where you are going to approach the second turn. This is where you should start to rate your horse (see figure 7.1). When your horse is completely turned and facing the second barrel, with his body in position to go to it, that's when you speed up. If your horse just has his nose around the barrel and you try to speed up, you will usually "arc off" or go too wide. This is a very common error for a lot of barrel racers.

After rounding the first barrel, when you're facing the spot on the ground where you're going to rate your horse and go into the second barrel, release your horse by pushing the reins forward on the horse's neck, and ask him to accelerate forward. Just like on the first barrel, never hold your horse's head to one side going into the second barrel. From the time your horse's tail clears the first barrel to half way across the pen, you should get in position for the second barrel. Before you have gone half way from the first to the second barrel, move your horse over gradually, change leads and be in position to rate and turn the second barrel.

Changing leads on your horse was fully discussed in Chapter 6, and should not be a problem once speed is introduced into the barrel pattern. By now it should come naturally for your horse to change leads, making it easier for him to turn the next barrel. However, if you feel that

After Martha leaves the first barrel, she looks ahead at the spot on the ground where she will be approaching the second barrel.

a) As Martha leaves the first barrel, she holds her horse straight and does not let him flex or arc his body.

b) After she leaves the first barrel, Martha boots the colt into a lope. Once she reaches the rate position, she will slow the colt down to walk or trot the second barrel.

your horse has not changed leads, your cue at this stage of the horse's training, would be a tug on the inside rein. (If you were to cue with the outside leg, the horse would possibly move over, toward the barrel, which would put him in a bad position to turn the barrel.) If you are having a problem with your horse changing leads between the first and second barrel, you can go around the first barrel, stop or slow your horse down and move him over with your inside leg. Repetition of this exercise will teach your horse to come around and get into position on his own. You should soon be able to move your speed back up and he will automatically change leads between barrels.

When you're going toward any barrel, never hold your horse's head sideways, by holding one rein tighter than the other. This will make him lose his ability to go forward. Do not hold any flex or arc in your horse's body while going to the barrel. Hold him straight until you reach the rate position and then get him in the "arc" position to turn. As you approach the second barrel, as long as you are in the correct position, never touch the horse's head. You want to "put him on his honor" or trust him, having him doing things on his own, as much as he can. This is also part of the "seasoning" process, which is getting a barrel horse to be fully trained and solid.

THE SECOND BARREL

The regulation barrel pattern is 60 feet from the timer to the first and second barrel, 90 feet between the first and second barrel and 105 feet to the third barrel. The second barrel is usually less of a problem than the first and third barrels because there is usually a shorter distance between the first and second barrels than there is between the second and third barrel. Of course, this depends on the size of the arena, as many barrel patterns are set up according to the size of the arena, rather than regulation size. However, most arenas are longer than they are wide.

Rate the same way on the second barrel as you did for the first barrel. It doesn't matter at what speed you are working, you want a horse to slow down (rate) and collect himself when he goes into the turn. That's because perfect "control" is when the horse can get around the barrel the fastest and leave the hardest. That is when he is gathered up and in the correct lead. Slowing down or rating in training teaches the horse to control forward motion.

If the horse moves out too wide in the pocket, use your legs and hands to move him in. If you want him to move left, push with your right leg and pull the reins to the left. If he has moved in toward the barrel too close, move him out. To move the horse to the right, away from the barrel, push him with your left leg. Use your right rein as the primary rein and the left rein to balance the horse, so he doesn't put his head to the outside. Just like the first barrel, rate by using your horse's front feet as the marker (see points #1 and #2 in figure 8.2).

To teach your horse to leave a barrel running hard, bring him around the barrel in position, release him when he's headed in the correct direction and then just let him run. If you do this, your horse will start leaving the barrel harder and harder. He'll soon understand that it's easy for

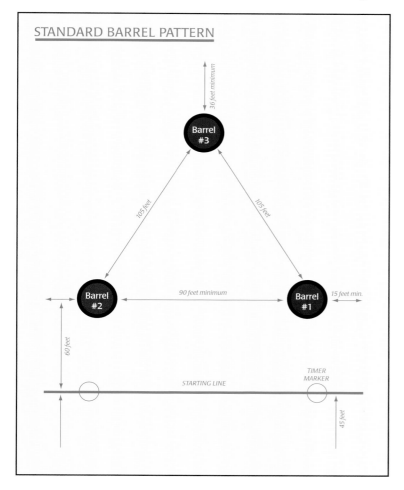

figure 8.3 *If an arena is not large enough to accommodate a standard barrel pattern, any size pattern may be used as long as the barrels are 15 feet from the fence.*

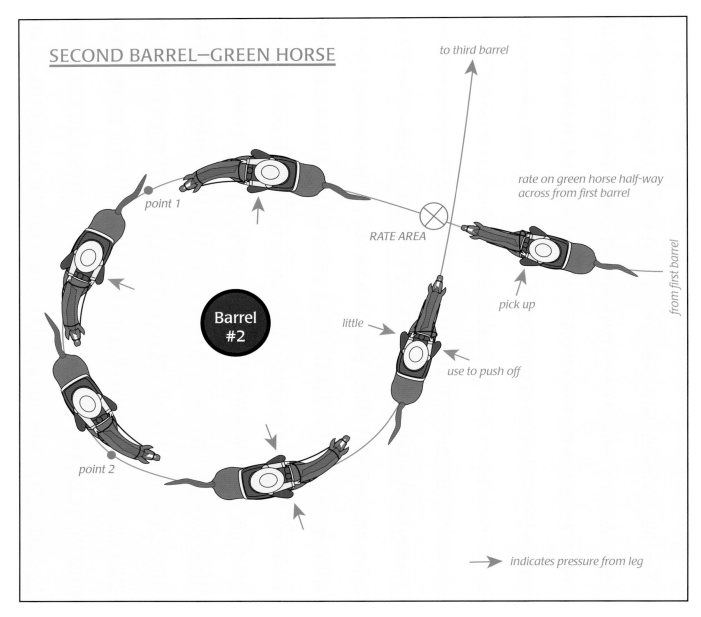

figure 8.4 On a green horse use both hands and inside leg pressure all the way around the second barrel to help the horse make a perfect pocket. Rate the horse at the point where the paths cross.

him and he'll automatically pick up speed when you release him. When you leave the second barrel, stay sitting down in the center of your saddle (not the back) until you are facing the point where you are going to go to the third barrel; then get forward into a jockey position. Your horse has already learned that this means for him to "go fast."

THE THIRD BARREL

The third and first barrels are the weakest for most everyone. When running at the first barrel, the horse usually has a great amount of speed built up and sometimes the rider doesn't rate him properly. Also, as we previously mentioned, sometimes the rider hasn't angled the horse right for an easy turn at the first barrel. However, on the third barrel, riders usually don't have problems rating it. The problem usually starts after the barrel has already been turned when they are leaving it. Most riders try to get their

INCREASING YOUR SPEED

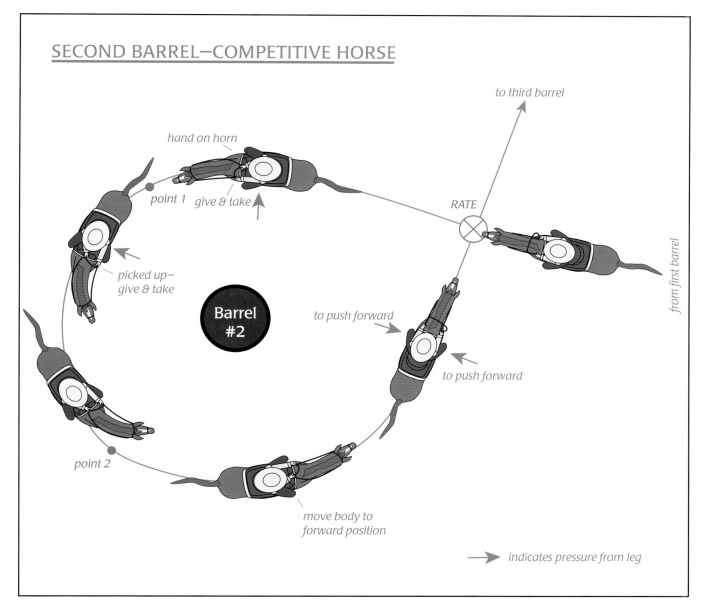

speed built up before they actually finish the turn. The horses, anticipating the pleasure of a fast run home, usually go along with them. A horse and rider have to be totally facing the finish line before they start to run out of the arena.

When going to the third barrel, head for the spot where you actually want to start your rate. As you approach the third barrel and get ready for your rate, you should be sitting in the center of your saddle and asking your horse to collect. The horse should rate and get into position (with a slight flex to his body). Give and take with your inside rein and nudge the horse with the calf of your inside leg to put some flex in his body to make the pocket.

The rate for the third barrel is accomplished the same as for the first and second barrel; however, on a competitive horse, you need to be really close to the third barrel before you actually rate. Come around the third barrel the same as the other barrels. Look at the ground ahead of you and

figure 8.5 On a competitive horse you can come into the second barrel with both hands on the reins, but you should take hold of the horn with your right hand just prior to point #1 where your turn starts which will help you balance your horse. After completing the turn at point #2, your body should be in a forward position to urge the horse forward and both hands can be back on the reins.

111

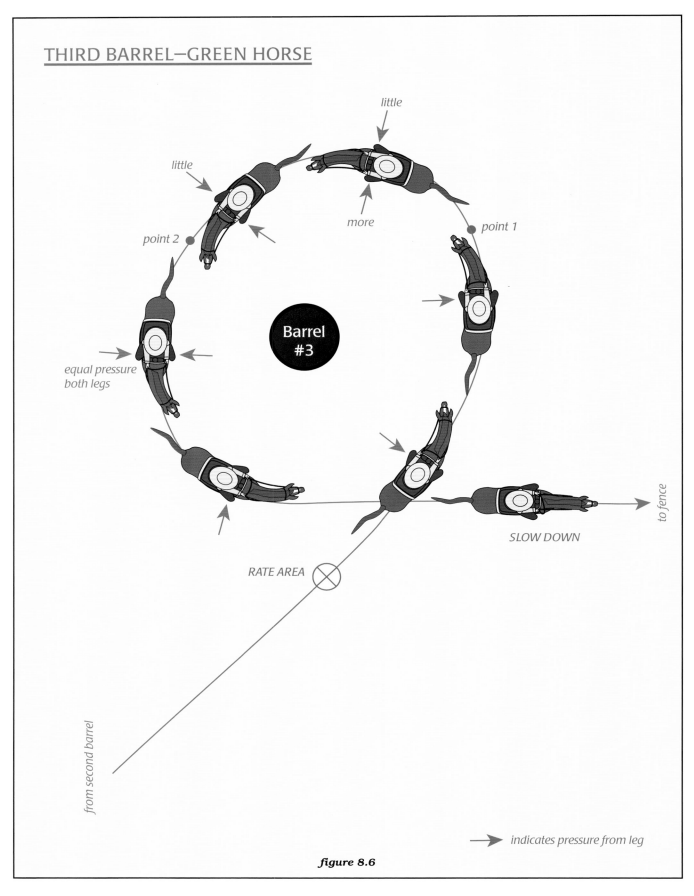

figure 8.6

INCREASING YOUR SPEED

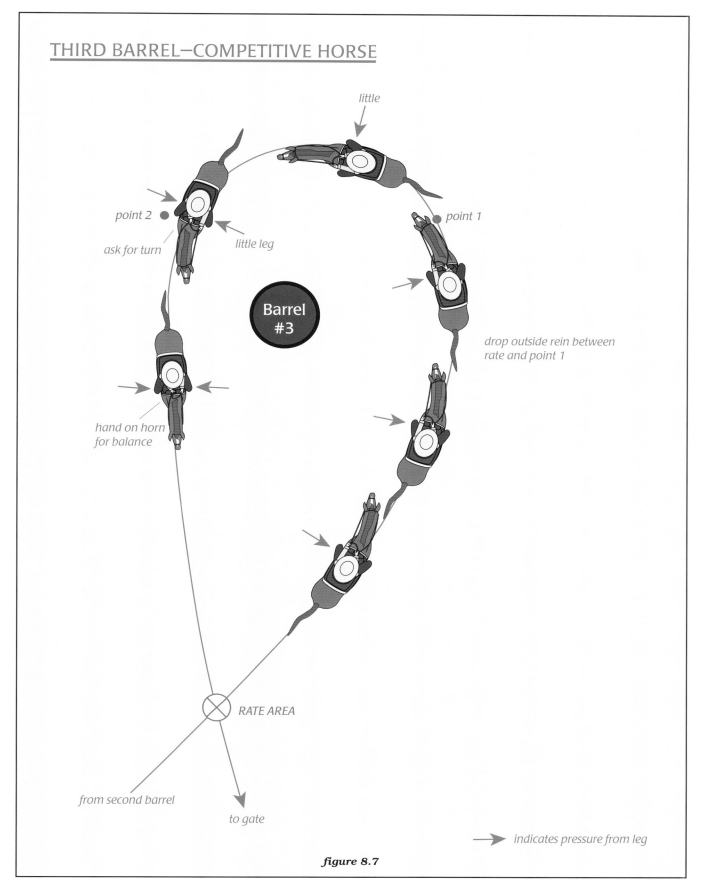

figure 8.7

(page 112) figure 8.6 With a green horse you make your complete pocket with boths hands on the reins. Use inside leg pressure all the way around the barrel, and once you have completed your turn at point #2, just keep turning and head for the arena fence, but slow down to a walk.

(page 113) figure 8.7 On a competitive horse you drop the outside rein and grab the horn with your right hand just before your turn starts at point #1. Once you have completed your turn at point #2, keep your outside hand on the horn to help you balance. In a competitive run, you would run straight out of the arena after turning the barrel.

a) Martha is making a practice run into the third barrel. Her right hand is on the horn and her left hand is holding the reins.
b) She still has her inside leg in the horse as she completes the turn at point #2.
c) Martha over-turns the barrel here and sits up straight in the saddle, which signals her horse to slow down. She will head toward the fence and walk her horse back to the starting line. Over-turning the third barrel prevents the horse from pulling out before he completes the third barrel.

as you finish rounding the barrel, look at the exit gate or the shortest distance to the finish line. You will still have light contact with the inside rein. Do not lose that contact before your horse has finished the turn, especially if your horse requires your help in completing the turn. If your horse doesn't come all the way around the third barrel and goes wide, or "blows" the barrel, you can usually fix this with a practice session at home.

When you are practicing, you should over-turn the third barrel and head for the fence at the side of the arena, slowing your horse down to a trot or a walk. Over-turning the barrel means to keep turning, even though it would ordinarily be time to go forward to the next barrel or run home. That way, your horse will not anticipate running out of the arena before he has finished his last turn. Usually a horse that anticipates his run home doesn't complete his turn. As a result, the horse will not wrap his body around the barrel, but rather will swing wide when coming out of the barrel and will lose time.

If your horse is used to you over-turning the barrel, he will listen to what you are signaling for him to do rather than anticipating running to the next barrel or home. If he listens to you and you let him go at the right place in the turn, his turn should be just right — not over-turned and not under-turned. Remember, however, that good judgment must be used by the rider to determine when to stop over-turning the barrel.

Also, when practicing, don't always run out of the arena. Even though a lot of time is made up on the long stretch home, your horse must not anticipate running before he has finished turning the third barrel. Save the hard run home for actual competition or when you are tuning your

horse. However, when you do run out of the arena, make sure your horse runs all the way to the finish line. At some arenas, the timer is set just inside the gate so if your horse starts "setting up" (slowing down in anticipation of stopping) before he reaches the timer, he will lose valuable time.

You have now moved your horse up through the barrel pattern, using the walk, trot, slow lope, fast lope, slow run and fast run. The next step is to make some exhibition runs at local jackpots and "season" the horse before you take him to his first real barrel race. In an exhibition run, you can pull the horse up and train him when he doesn't handle the speed properly. Jackpots are usually inexpensive, with not a lot of money at stake. They simulate actual show conditions, without the high cost.

How you regulate speed in your horse will really depend on the horse. If you have a four-year-old that is running barrels and is highly competitive, use good judgment on his attitude. If he's real racy and hyper when you leave home, keep him real quiet and do a lot of slow work on him. When you get to the jackpot, just exhibition him (make a practice run) by walking or trotting, or going no faster than a slow, controlled lope. On the other hand, if your horse is a laid-back or non-aggressive horse, compete on him a little. Enter some jackpots when you feel he needs sharpening up. A competitive run will free him up mentally (make him want to run).

We rarely take our four-year-olds to compete in barrel races at rodeos because it is just too stressful on them. It's too much pressure to put on them. The crowd is noisy and it generally scares young horses. Then, too, there is pressure to win due to the money that is up. Also, at a rodeo, the ground conditions usually aren't the best. We want our horses more mentally mature than most four-year-olds are before they compete in a rodeo. They need to get accustomed to noisy bands, waving flags, bright lights and all the other scary things in the world of rodeo barrel racing.

Also, we don't take any of our horses to futurities until they are four years old. Before that, we feel that they are too immature, mentally and physically, to be able to take that type of pressure and speed. Once in awhile, you will see someone take a three-year-old to a futurity, but it's not often that you will see one excel. That's usually because a three-year-old is mentally and physically a year behind the four-year-old. There's a great spread in mentality and maturity between the two ages. Due to the large amount of money they pay out, the futurities have gotten so tough that you've got to have the advantage of an older horse with more experience. You want a horse so mentally and physically ready that he's almost like a seasoned veteran by the time he gets to the futurity.

"THE TIME IT TAKES TO MAKE A FINISHED BARREL HORSE IS NOT A FACTOR."

9

Exercises to Improve Your Run

There are a variety of exercises and drills to teach a horse to be soft or supple. These exercises should be done before a horse is started on barrels, while the horse is being trained and also with a horse that is totally trained. These are drills that reinforce and enhance all of the education that has already been taught to your horse ... and to you. They help keep a horse supple in the body and teach him to get in position around the barrels; yet they won't stress him mentally or physically.

However, for any of these drills to work, a rider must be in time with the horse. Timing is doing the right thing at the right time. If you do a maneuver at the wrong time, it's the same as doing something wrong. For example, if you rate a horse perfectly, but it's too early, and then you "run him through the rate position" to get to the barrel, you have actually taught the horse to pass, or run through the rate position of the barrel, rather than slow down and rate.

Education, which leads to communication, along with proper timing and repetition added in, equals the finished horse. "Perfect" repetition is needed to make the finished barrel horse. The period of time that it takes to make a finished barrel horse is not a factor in the equation. It varies with each horse. We have come up with some drills that will keep a horse supple as well as help him change leads and hold the correct lead in a turn around a barrel. These drills are called "big circles," "loops," "half-arounds," "spirals" and "four-cornered circles."

However, one word of caution. While you want a horse to be supple, be careful that you don't over-bend or over-flex him. That's worse than having him too stiff. A horse that bends too much loses his ability to go forward in position and to hold that position. He also loses his ability

(overleaf) When doing exercises, keep your horse's whole body—hips, ribs, shoulders, neck and nose—in a slight arc. These exercises will keep your horse supple and allow him to bend more easily.

to protect himself because he's bent around like a noodle in a circle. He's just like a person trying to run with his torso twisted sideways. He can't run very fast without injuring himself.

BIG CIRCLES

Big circles and loops (see figures 9.1 and 9.2) are useful drills for training barrel prospects or for tuning finished horses. These drills are especially ideal for a horse which does not complete a barrel turn properly. The "big circle" drill (figure 9.1) teaches a horse how to (1) hold his leads, (2) keep his feet under him, (3) pick up his shoulder, (4) follow his nose, (5) drive with his rear-end and (6) maintain his forward motion.

With a young horse, you can perform these drills at a long trot or a slow lope. On an older horse, which just needs tuning, or for an older horse with a problem, you can do the drills with quite a bit of speed — say around 70 to 90 percent of full speed. To a certain degree, an older horse can be corrected while working slowly, but he must eventually be corrected with speed, because with speed, most horses go back to their old habits. The correction, therefore, must be accomplished at all speeds. We do the first drill, "big circles," followed by the second drill, "loops," and they are always done in that order.

To do the "big circles" drill, go to the first barrel, staying approximately 10 feet from it, but turn all the way around it, with a very symmetrical circle. Long trot and lope your circles around the barrel two or three times, holding the size of the circle that you want. If a horse is slightly dropping in toward the barrel, we would move him out a little bit, approximately 15 feet from the barrel. If we really need a horse to move out because he's really wanting to "drop" into the barrel, trapping us, we would use the 20-foot circle. When a horse "drops in" toward the barrel, he is just overanxious to turn the barrel. It must be reinforced in the horse's mind that he has to honor a certain position while going around the barrel.

While going around the barrel, the horse must be in the proper lead, and his whole body (hip, ribs, shoulder, neck and nose) in a slight arc. A horse that is more supple can bend a little more, while a horse that is a little stiffer can bend less. Some horses are able to bend quite a bit, but you never want them bent too much because they lose their athletic ability. How do you arc your horse's body? On a young, uneducated horse, tugging on the inside rein tips the nose in slightly. Pressure on the calf of the inside leg flexes the rib cage.

For a right-hand barrel, your right hand should go across the horse's neck indirectly with the inside rein to move the shoulder and rib out. The rein moves the horse's shoulder, neck and head. Your leg moves the horse's rib cage and shoulder. It takes both to move the horse over. Slight tension on the outside rein collects the body and holds the horse in position so that he doesn't move too far out from the barrel.

(opposite) figure 9.1 Long trot or lope around each barrel two or three times, approximately 15 feet from the barrel. If the horse wants to drop into the barrel, circle 20 feet from the barrel, but make your horse honor whatever pocket you choose. After circling each barrel, stop your horse and let him relax and think about what he just did.

BIG CIRCLES

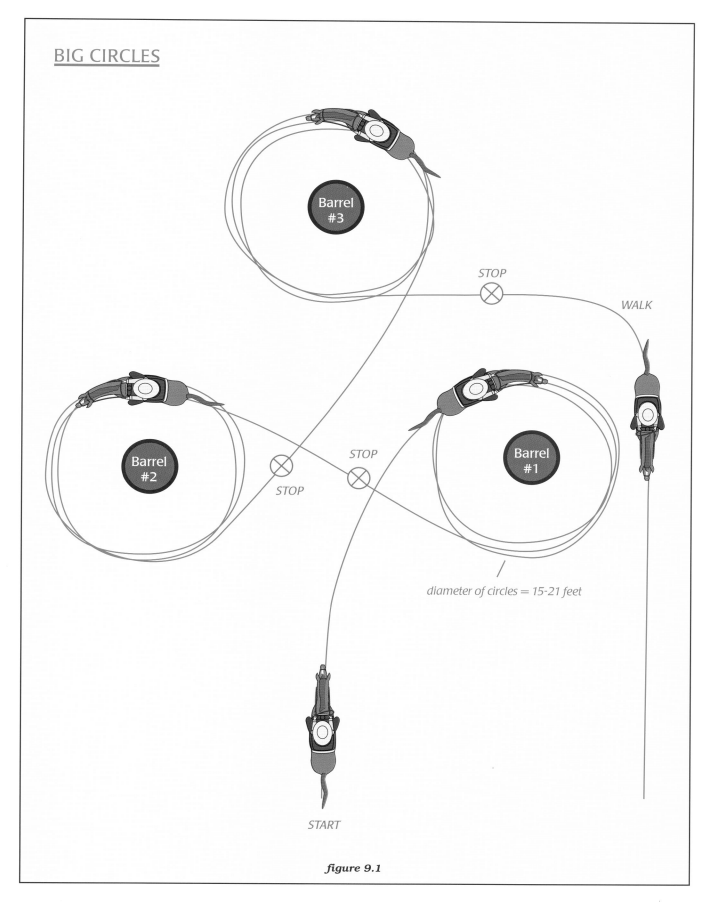

figure 9.1

Ed teaches horses which are tense or "on the muscle" to relax with circles and loops. A horse needs to be soft in the body and mind for him to retain what you are trying to teach him. If he is afraid or tense, he won't learn.

If the horse is moving out too much, with too much bend, bring the outside or "balance" rein in toward the horse. If he needs to go forward more, ride him with the calves of both legs. If he needs to slow down more, bring the inside rein directly back to his right shoulder in a "give-and-take" motion to slow him down. After habits are formed in the horse's mind, control is needed less and less by the rider as the horse is learning to do more and more himself.

Then stop, let the horse relax, settle and digest the thought of what he has just done. Anytime you do a drill, whether it's walking or running, you want a horse soft in the body and the mind, because that's the only way he will retain what he's learned. If a horse is tense, he won't retain anything and you'll also probably create new problems in the process. After the horse feels settled under you, get back in the correct lead (if you are loping) and go to the second barrel. Move him over in preparation for his approach for the second barrel. You do this with your legs and hands, by putting pressure on him with the calf of your outside leg and moving both hands toward the direction you are going.

If the horse is too far out, bring your outside leg and your outside hand in toward him. If he's out in the rear end, use the outside leg to contain him, plus the outside hand. If he's just out in the shoulder, only use the outside hand, bringing it in toward the horse. Go around the barrel at the same speed, 10 to 15 feet from the barrel, until the horse has gone around it two or three times properly. Leave the second barrel, stop and let the horse digest the thought of what he's done. After he's settled, go to the third barrel and repeat the process, again moving the horse into position.

When you are practicing, we suggest that half of the time you should over-turn the third barrel, then go to the fence and come down the fence on the back side of the first barrel. Over-turn means to turn the barrel to the degree of leaving the third barrel, then going in the direction behind the first barrel, along the fence and back to the starting line (see figure 9.1). This is done half of the time because a horse also needs to learn to come straight out of the arena. By over-turning the barrel only half of the time, the horse will be waiting for the rider's cue signaling him as to what he should do — over-turn the barrel or come out of the arena. This drill for over-turning the barrel is done because many people try to go to the finish line before the third barrel is completely turned.

If you over-turn a barrel enough, the horse will start thinking he should over-turn the barrel. That way, he won't think he should run out of the arena before he turns the barrel completely. You want the horse to do 95 percent of the thinking if he's the type of horse that will allow you to let him. Over-turning a barrel will let him do that. However, some people can over-turn too much and a horse won't leave a barrel with the speed he needs to, because he's thinking over-turn. That's why we suggest over-turning the barrel only 50 percent of the time.

Common sense must be used, however, depending on the horse. If a horse over-turns and goes to the over-turn position too often, then only over-turn him 20 percent of the time. If he doesn't finish the third barrel, then over-turn him 80 to 90 percent of the time. All the drills and all the training call for an educated rider to use proper judgment. This is only effective if your timing is with the horse. You can't do any of these drills properly if your timing is off. Just like you can't compete with bad timing. This drill is only used on the third barrel because that is the only barrel that a horse has to leave and run straight out of the arena.

LOOPS

For loops (figure 9.2), repeat your circles from the previous drill at the first barrel, long trotting or loping; or if you have an older, seasoned horse, use some speed. After you do circles around the barrels, add to that exercise by doing loops around pylons. Come out, away from the circle, directly behind the first barrel and do loops around a pylon, around 15 to 18 feet from it. Go around the pylon two or three times, just like your circles, but these are designated as loops because of where you are and what you are going around — a pylon rather than a barrel.

A loop does all the same things that circles do, but it also forces a horse to completely turn the barrel. Plus, it makes a horse listen to his rider. If done properly, this drill does not make a horse want to leave the first barrel and go to the loop position. But you can overdo loops and create a

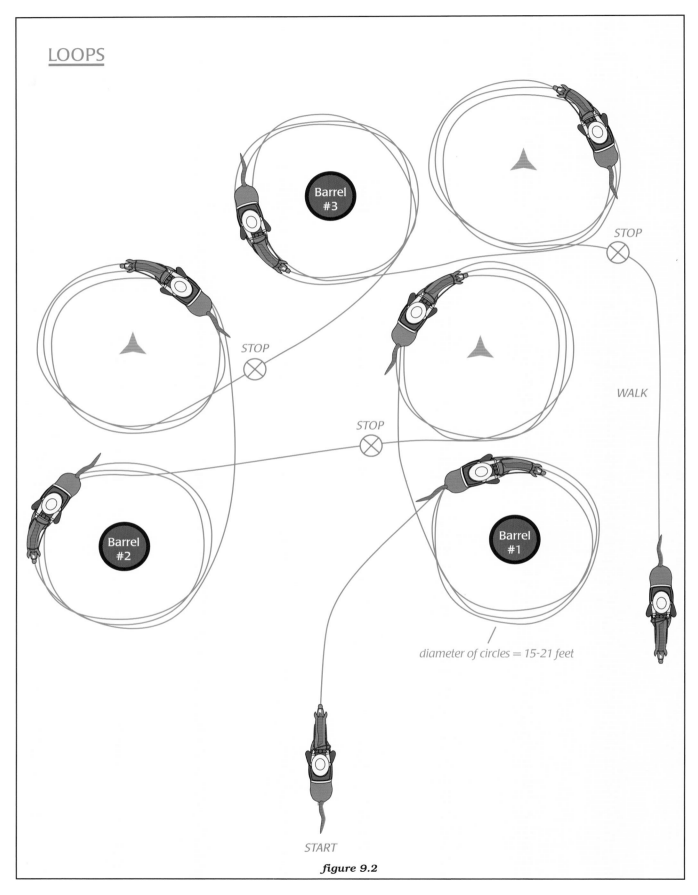

figure 9.2

problem for yourself. Over-educating a horse on the loops makes him think he is trained to do loops at high speed when competing instead of just using the loop as a training drill. By using good judgment, however, you should determine when the drill has accomplished what you are trying to teach the horse. If done properly, the loops exercise reinforces everything a horse does with circles.

The place for the first loop is between the first and third barrel, a little to the outside of a line drawn from the first to the third barrel. The loop is where you can put pressure on a horse if he is not giving you his full attention. This is where you can scold a horse for hitting a barrel or for not rating and, thereby, going by a barrel. You can put "heat" on a horse (asking him for speed by urging him forward with your legs or tapping him on the hind end) in the loop, mentally and physically, then come back to the barrel, go around it and let the horse relax.

You can teach a horse to really finish his turn by over-turning a barrel. Make sure your horse holds the proper lead, drives with his rear end, and keeps his body in position all the way around the turn. Anytime you need to put set or rate in a horse, do it BEFORE you approach the loops and BEFORE you approach the barrel. When you get through the loop at the first barrel, stop. That is where you can teach the horse to change leads as he leaves the first barrel. Teach him to get over and into position for turning the second barrel. Changing leads requires collection, or slowing of the horse's forward motion, and the use of outside leg pressure that is applied by the rider. The outside hand can also be used as needed. Changing leads was fully described in Chapter 6.

Once you've finished doing loops on the first barrel, go to the second barrel, turn it and then go into the loop. The loop on the second barrel is straight up from the second barrel, to the left of a line drawn from the second to the third barrel. The loop for the third barrel or pylon is located directly out from the third barrel, between the third barrel and the fence on the first-barrel side of the arena.

When working at slow to medium speed, you should finish the third barrel by turning it in the direction of the fence, then come down the fence to the finish line. By overemphasizing the finish on the third barrel, you won't make the mistake of trying to get out of the arena before you are actually facing the finish line. When we finish the third barrel at home, we pull up and walk or trot out. It's really easy to get a horse to run out of the pen. Almost anybody can do that. We want a horse quiet and under control.

We work our colts a couple of times a week with big circles and loops, just to keep them relaxed, yet sharp at the barrel pattern. The big circles and loops actually simulate turns around the barrels. They let the horses relax and let the rider practice the positions he needs to put his horse in. These exercises are important because they let a horse and rider practice the competitive situation, without the stress involved in it.

(opposite) figure 9.2 The "Loops" exercise will help your horse relax, make him listen to your cues, and encourage him to completely turn the barrel. Pylons are placed above the first and second barrels and to the side of the third barrel.

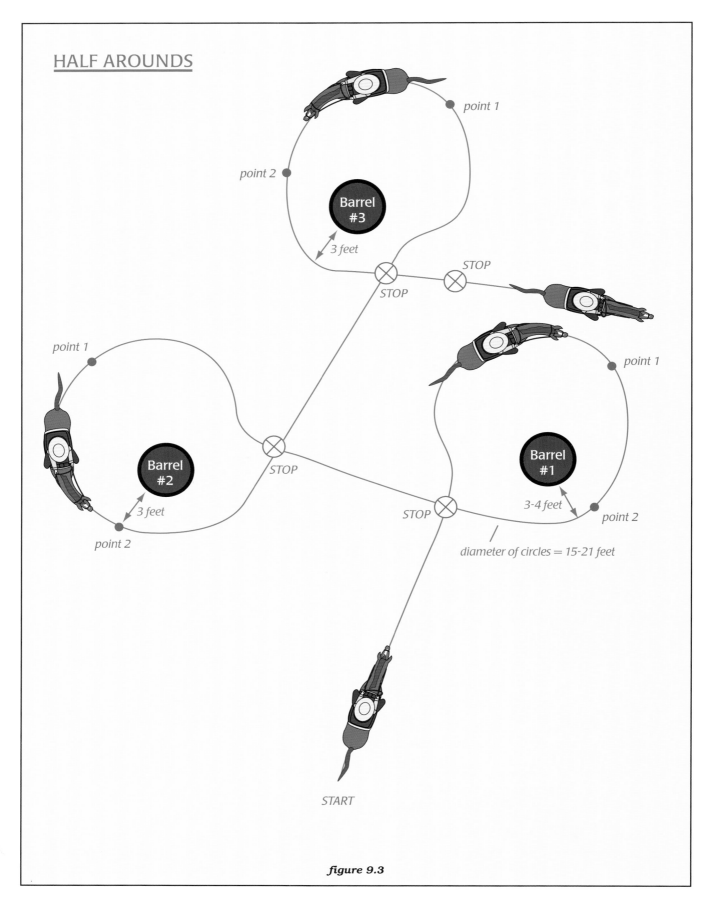
figure 9.3

EXERCISES TO IMPROVE YOUR RUN

HALF-AROUNDS

Half-arounds teach a horse to hold his pocket or not "blow" (leave) his pocket at the barrel. This is a good exercise for a horse that wants to move out of his pocket or wants to hit a barrel going in or coming out. They're great if your horse has too much set too early, or hugs a barrel too tightly, where he can't run around it and still keep his forward motion. Unlike the other two drills, this drill is for correcting problems. It is not a drill that we do all the time on any horse.

To do this drill, approach the barrel at the regular position and stop or slow down (rate) your horse approximately 12 to 15 feet from the first barrel. Collect the horse with your reins and move him to the left by using pressure from the calf of your right leg. Exaggerate the pocket from 15 to 20 feet on the back side of the barrel between point #1 and point #2. Then drop in to your regular turn position and finish the rest of the barrel pattern with regular turns, which are approximately two or three feet off the barrel. This simply overemphasizes the pocket (see figure 9.3).

As you leave the first barrel, stop, let your horse digest what you've just done and relax. Move over and go to the next barrel as you would normally. Then drop out approximately 15 to 20 feet from the barrel, making your overemphasized pocket. Half way around, between point #1 and point #2, drop back in and finish two or three feet from the barrel. Come back around the barrel if you need to repeat it and keep overemphasizing your pocket, then drop back in to finish the turn. As you come off the second barrel, stop your horse and let him relax. Lope to the third barrel, stop, move out, overemphasize the pocket, drop back in and do a regular turn.

At first, do this exercise while long trotting and loping, but with a problem horse, or an older horse, you have to have speed. As we said before, 90 percent of the time, you can't correct an older, seasoned horse going slowly. You usually have to use speed and be in a competitive situation. Therefore, he can't be totally corrected at home. Half-arounds make the horse respond to the position you want him in, plus they train him to relax. He must first respond to what you want him to do. Secondly, get your horse to form a habit of being in the correct position so that he knows that's where he's supposed to be.

SPIRALS

Spirals are a good drill to do before you ever get to the barrel pattern, but can also be used throughout the horse's training and after he is a solid barrel horse, just to reinforce his training. It helps you and your horse prepare for the barrel pattern because it lets you move the horse in from the outside to the center of the circle and back out. This sets the stage for total communication when you and your horse get to the barrel pattern (see figure 9.4).

Start loping a 90-foot circle and drop down, spiraling inward until you're down to a 15-foot circle. Keep loping and spiral right on back out. Do it in both directions. While

a) The "half-around" exercise — good for horses that "hug" a barrel too closely — exaggerates the pocket on the back side of the barrel. First, Ed collects his horse before coming into the barrel and moves him over using pressure from his right leg.
b) Once Ed releases the pressure from his inside leg, the horse moves into the barrel and attempts to "hug" it, but was still in a good enough position to run around the barrel.
c) Ed completes his turn, making it a nice run.

figure 9.3 (opposite) "Half Arounds" is a good exercise for correcting a horse with too much early set or a horse which hugs a barrel too tightly.

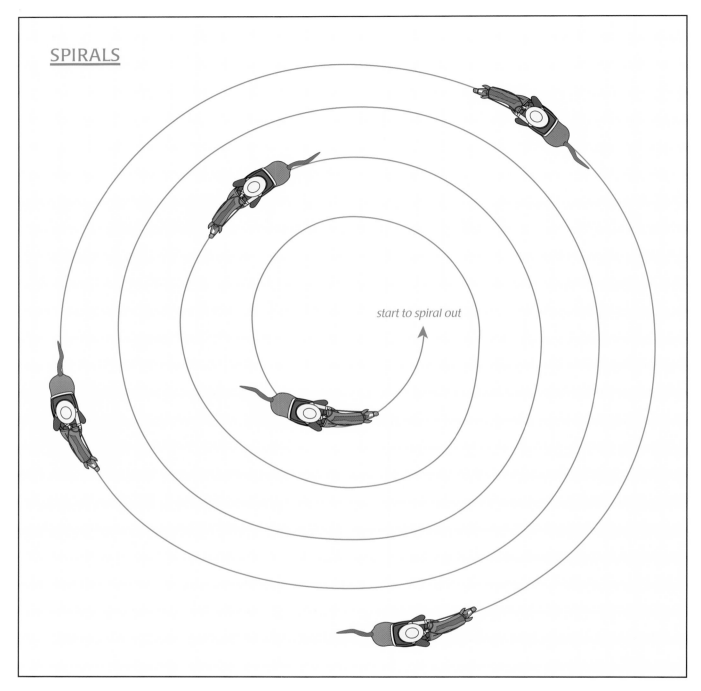

figure 9.4 The "Spirals" drill reinforces a horse's training by letting you move the horse in from the outside to the center of the circle and back out. Start with a 90-foot diameter circle and spiral in to a 15-foot diameter circle.

performing the spirals, ride the horse forward into the bit by using leg pressure. Hold the horse in position, giving and taking with the reins. As we've mentioned before, never hold the horse solidly; you always give and take. Keep your horse soft, take hold of him and let go, take hold, let go. Your horse must stay soft — soft in the mind, to respond to anything you ask him to do, as well as his body.

FOUR-CORNERED CIRCLES

The four-cornered circle drill (figure 9.5) is for horses of any age or particular level of training. Start loping a 90-foot circle, which is your main circle. When you reach a

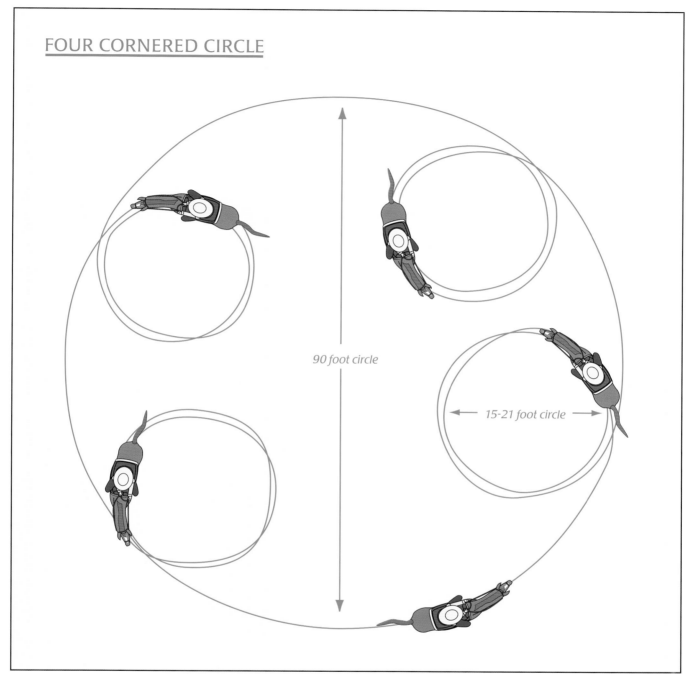

corner, make a 15-20 foot circle, then lope right back out into the main circle. Lope to the next corner, lope a 15-20 foot circle, then keep loping the main circle. This exercise teaches a horse to hold his leads when he goes in and makes a barrel turn and when he comes out of it. It teaches him to hold his body position and travel forward before he ever gets to the barrel pattern. Do this drill equally to the left and to the right.

figure 9.5 The "Four-cornered Circle," which teaches a horse to hold his leads, begins with a 90-foot circle. When the horse reaches each of the "four corners" of the circle, lope several 15-20 foot circles, then continue on the main circle again.

PASTURE LOOPS

We have an $^8/_{10}$ of a mile track around the pasture. That's where we warm up. We lope around it on the left

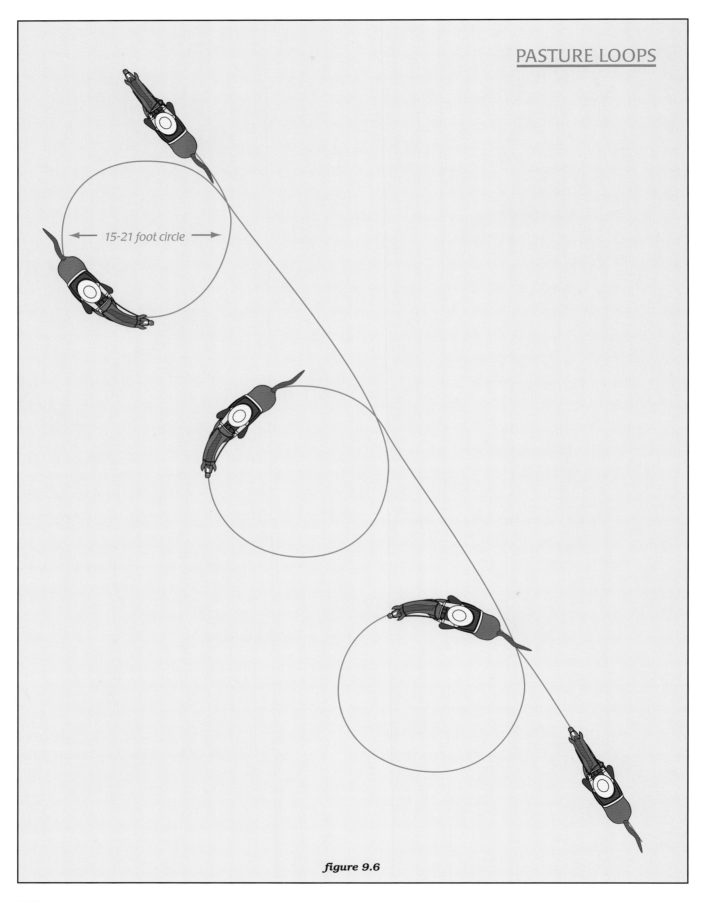

figure 9.6

lead, then make a turn, two to four times the size of a barrel turn, to the left and then lope right back around the pasture. Then we make another little turn. This must be done to the right and left equally. This drill teaches a horse collection, response and leads.

FREEING UP A HORSE TO RUN OUT

When we want to free up or "loosen up" a horse, sprinting (running fast for a short distance on a straight-away) doesn't always help. Freeing up a horse can be done on the barrel pattern and you can use large circles and the half-arounds. It may sound crazy, but they also put set and turn in a horse. The following exercise teaches proper position.

If you lope to the barrel, stop and lope the big circles, which puts "set" in a horse. When you go up to the barrel and never slow down, freely loping your big circles, you put a little bit more "free" in the horse and really loosen him up, giving him the desire to move out and run. If your horse doesn't "free up" that way, let him work the pattern at his own speed. Then gently hold his head while you are loping around the barrel, and tap him lightly with a bat or an over-and-under. This will free up a horse better than turning him loose.

If you have a horse that won't free up running out of an arena, find a place where it is safe to run out of the pen and run another 50 to 60 yards after you have crossed the finish line. That way, you can put a lot of run in a horse's mind if he previously wanted to set up before he reached the finish line. If used with good judgment, the above exercises can help you keep your horse supple if he is an older, solid horse, or teach him how to get in position for each barrel if he is a young horse — and without stressing him mentally or physically.

(opposite) figure 9.6 To perform "Pasture Loops," lope a straight line, making 15-20 foot circles on one side of the line. This drill must be done equally to the right and the left.

"YOU MUST USE GOOD JUDGMENT IN CONDITIONING."

10

Conditioning the Athlete

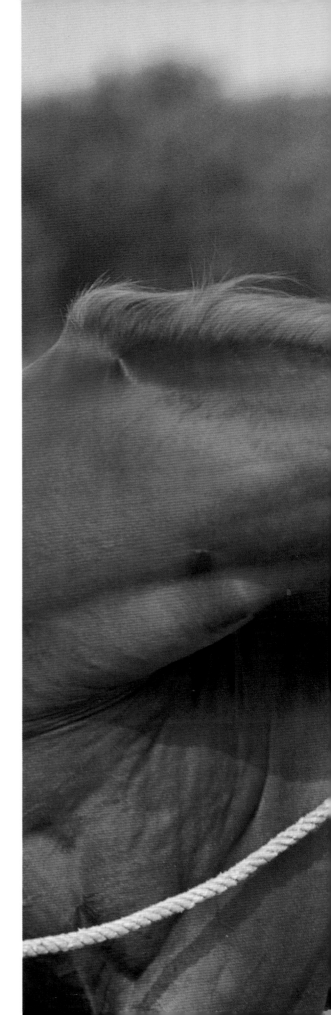

Your horse should be in good condition to run barrels. If he's not fit and you work him on the barrels, he might relate the barrels to getting sore. If that's the case, he won't like what he's doing for very long. You must use good judgment in conditioning. You can tell, to a certain degree, if a horse is in good condition just by looking at his coat because it's fairly easy to determine whether it's shiny or dull. But you can be fooled. A horse with a shiny coat is usually healthy, but if he has a big hay belly, you can be assured he's not in condition.

You can also observe the horse's condition when you are riding him. If he's not getting enough oxygen, he will be breathing extremely heavy, and he must be walked until he has regained his oxygen supply. You can also monitor how hot he is by the amount of sweat he has on his body, as well as his body temperature. If he is hot, he must be walked and cooled out slowly. You should also check the elasticity of the horse's tendons, ligaments and suspensory system, as well as his muscle tone and suppleness. When checking for elasticity, make sure there isn't any swelling or puffiness around the joints or down the back of the legs.

A horse with good muscle tone has smooth and supple muscles, with no visible or palpable muscle spasms or soreness. You need a good, smooth muscle that is not too soft and flabby, but it shouldn't be extra hard. Real hard muscle has too much bulk and makes the horse susceptible to injury. It also keeps the horse from being quite as athletic as a horse with good, smooth, supple muscles. A horse should be built up gradually to the distance you want him to go. Conditioning is a combination of distance and the time it takes to go that distance. A horse should do part of his conditioning by long trotting and the other part by short (slow) loping.

(overleaf) Stretching is an important part of conditioning and should be done daily. Martha uses treats to encourage her horses to stretch.

On a horse that wants to lope fast, this is accomplished by softly seesawing the reins, alternating pressure on one rein and then the other. If the horse is moving forward with no leg pressure from the rider, he is an aggressive horse. The less aggressive horse must be asked to move forward with leg contact or pressure, which will result in a collected lope if the rider seesaws softly on the reins.

Sprinting is not often a part of conditioning, especially if you are making quite a few runs on your horse each week. A combination of long trotting and short loping is best because when you come to a barrel, you want your horse to shorten his stride and be quick around the barrels. However, in between those barrels, you want him to reach out and lengthen his stride. Long trotting makes a horse reach and take hold of the ground. Short loping conditions him to shorten his stride and be more collected.

You might not be able to do these exercises 50/50 on every horse. A horse with a Thoroughbred-like (long) stride won't need much long trotting to lengthen his stride. You would probably do 75 to 80 percent of his conditioning short loping to shorten him up and the remaining 20 to 25 percent long trotting. It is the same with a choppy horse. You would do 75 to 80 percent of his exercise by long trotting him to stretch him out and only 20 to 25 percent short loping.

To condition a horse, ride him for six to eight miles, three days a week every other day. During the three days in between, condition the horse easily by just riding him out in the pasture. Then, once a week, turn him out and let him relax. The way you condition is by monitoring your distance and the speed with which you go that distance.

WHERE TO CONDITION YOUR HORSE

You should locate an arena that is soft, safe and long enough to exercise your horse. But don't exercise only inside an arena. We have a track that is eight-tenths of a mile long and we condition our horses on that track. But you don't need a track to condition your horses at home. You can measure the distance where you usually ride with your pickup, car or motorcycle.

For a horse that isn't in condition, start out by going three-tenths to half of a mile, trotting and loping. Then every four or five days, move up by two-tenths of a mile. It takes a while to work up to six or eight miles. It takes a really long time to get a horse in top physical condition. It takes an average time of 90 days before your horse reaches his upper levels of conditioning and six months before reaching peak condition.

Anytime your horse is out of air, or is too hot, you need to slow down and walk. If it's high noon in the summer, stop and let your horse walk and cool out often. If it's December and 40 degrees, he can probably go a longer distance before you have trouble with overheating.

When your horse is hot, never give him all the water he wants to drink. Let him drink one swallow of water every two minutes or so during the cooling process. When he refuses water, he is cooled out. Never just stop and tie him up. You should walk him slowly and let him get back to where he has reached his normal temperature and respiration.

HILL WORK

Anytime you run barrels on a horse, he uses his muscles by turning and pushing in a different way than the Good Lord meant for him to. Consequently, we do a lot of circles on the side of a hill to help teach the horse to hold his position, balance and pay attention to his footing. There are several reasons to use an incline to lope on: Hill work on your horse is great for conditioning and is a body-building process for the horse. We don't have any hills on our property, but we have a stock tank with a dirt dam on it. We lope large circles both ways on the side of the dam or the hill. It teaches a horse to use himself a little more. Also, he's more aware of where he's putting his feet and where his body is.

When you're loping circles on the side of a hill, be careful of the terrain. Our stock tank dam, like most, is not really level. It has some washed-out places and a few armadillo holes in it. If a horse hangs up, stumbles or falls in a hole by not being attentive, he can go down and roll over on you. To get your horse used to the terrain, walk your horse for the first few circles going both ways. Then trot a few circles. When your horse starts thinking about where his feet are, you can lope.

Don't do any small circles for this exercise. Your circles should be from 30 to 60 feet. If we have access to a hill, we do some drills going up the hills, like the loops and spirals. If you're lucky enough to live where you can gather cattle, goats or sheep, and work up and down the hills, you'll be doing the best conditioning for a horse, physically and mentally.

Doing a lot of circles on the side of a hill will help teach your horse how to hold his position, balance himself, and pay attention to his footing.

Horse treats are an easy and effective way to get your horse to stretch important neck muscles.

STRETCHING

Stretching is just as important as conditioning and you should stretch your horse on a daily basis. We get a horse to stretch himself by using hay or horse treats as incentives. Horse treats, which are usually a combination of molasses, crushed corn, alfalfa and sugar, are sold under several brand names by several companies. You can find them at most local feed stores.

Feeding a horse treats really goes against what many of us learned earlier. We were always taught never to feed a horse out of our hands. But we feed all our horses horse treats as a reward or to encourage them to stretch. Horse treats are also good to help with the bonding process between horse and human. There are some horses, especially those that come from the track, that are totally negative about humans. They won't bond with people; they don't even like people. Such horses have to be taught to perceive humans in a different light. They have to learn to like humans. Treats can be a good tool in those cases.

You can stretch your horse all the way from his nose to his hindquarters, just by placing horse treats so the horse has to stretch to reach them. A forced stretch is not good. That's just like somebody stretching your leg when you're lying on the floor. They can't really tell how far to go by feel. You want a horse's nose to go behind his shoulder to the girth area and to the last rib on both sides.

Another good stretch is to have the horse touch his nose to his chest and then go between his front legs, without him buckling his knees. You can do this by placing the treat on the horse's chest and then between his front legs toward his stomach. He'll have to reach underneath himself to get it.

Your horse needs to be stretched in the morning and evening when you feed him and before and after you ride him. Stretching is imperative and may be as important as conditioning. If you have a horse that is really in condition and his muscles are hard and tight, he needs to be stretched so he doesn't injure himself during exertion.

It is important to stretch your horse before you run the barrel pattern and after you are through running. Stretching keeps the body energies flowing and helps rid the muscles of spasms. When you stretch, you let the "chi" flow, which is the Chinese term for body energies. There is more about this in Chapter 14.

Eliminating soreness is another benefit of stretching and can help keep major injuries from happening. If your horse has a pulled muscle and you don't work on that muscle, you're going to have other damage, more than likely to the ligaments and tendons. A pulled muscle can even start a calcium deposit. Inflammation and pain are the body's way of telling the mind not to use that location because it is impaired.

WARMING UP YOUR HORSE

Your horse needs to be warmed up before you work him, whether it be for practice or running barrels in competition. Obviously warming up for a slow work does not have to be as vigorous as warming up for competition. When warming up a horse, you should monitor what you're doing and don't overdo it. For example, if you're trying to ride the edge off of a rank horse, don't ride him until he's too tired or grouchy to compete. Ride your horse until he is mentally ready to compete, spending a minimum of 15 to 20 minutes warming up. Keep in mind, however, that each horse is different. Some may only need to be ridden a minimum amount of time and some a long time in order to have them focused and ready to work.

When you come to the arena gate to run barrels, you don't want your horse to be out of oxygen from an excessive high-speed warm-up. If you've gotten him too hot, you need to walk him before you run him or work him. You want him level and focused. Never get a horse too hot while warming him up. If it's cold out, you may have to lope your horse about 20 minutes until he feels like he's loose. If your horse is older and needs more work than that, do what it takes until he's supple.

COOLING OUT THE BARREL HORSE

Always wash all the sweat and dirt off your horse after you ride him when the weather allows. When you're all sweaty and dirty, you don't feel good the next morning unless you wash all the sweat and dirt off. Even if your horse rolls and you don't have time to wait for him to dry off before you turn him loose, dirt's easy to brush off. Sweat is hard to brush off. Hosing or rinsing a horse with cool water is probably the best thing you can use to physically cool out any horse. We try to remove heat from our horses as quickly as possible when they come out of the arena. Obviously, if it's a cold winter day, we don't worry about getting water on them.

As soon as possible, we also like to get the saddle off and wash the horse, preferably with a body wash. The body wash can consist of a mixture of water and another additive suggested by the vet, or you can purchase one already made up over the counter at a tack store. Wash the horse with this for the removal of heat in the horse. It will also help the horse be less body sore and tired from a hard work. Not only does a body wash cool a horse quicker, but it pulls out soreness and helps circulate the blood. It also helps lower the horse's body temperature and flushes out the system by dissipating the heat as soon as possible, which is your goal in cooling a horse.

If it's cold, you can't give your horse a body wash, but you can wash under his belly and neck where the most blood flow is. You can also wash down his legs to cool them off and keep them tight. Anytime that it is raining and damp, you need a sheet to put on your horse until you get him cooled out a little bit. Then you're going to have to pull that sheet off and put a dry one on him. It takes a lot of work and it may be a headache, but it's really important if you want to win and keep winning.

"PERFECT PRACTICE MAKES PERFECT."

11

Mental Preparation

To be a champion, you must think you are a champion! Always remember that you can defeat yourself in a barrel race by not preparing mentally. One of the main things you can do to prepare yourself mentally to be a winner is to control your emotions. Try not to be real high or real low. If you're real high, wound up and your adrenaline is flowing, you're not going to think as sharply. The same if you're feeling real low and unsure of yourself. You should stay level.

In competition, you need to think all the time. You should be sharp in your thought processes, aware of everything you do when you go through the barrel pattern and able to react positively in all situations. Anytime you make an error, don't dwell on it. If you mess up on the first barrel, erase it from you mind and just do your best on the other two barrels. Also, after your run is over, don't dwell on your mistakes for three or four weeks. Use them to learn.

There are several things you can do to help you prepare yourself mentally. Deep-breathing exercises help. But when you breathe deeply, don't just breathe down to your stomach. Breathe all the way down to your hips. Breathe slowly... and then you'll be able to think better. Confidence is also a major factor in barrel racing. You gain confidence through knowledge, practice and repetition. The more confidence you have, the greater your ability will be.

Going over your runs can also help you with your mental preparation. When you're at home, get by yourself where there's no television, people or noise. For five to 10 minutes every night, go over the runs you made that day in your mind — in slow motion. But don't go too fast because you will miss steps. Go through your run, picturing you and your horse making a perfect run. That perfect run will refresh your memory.

(overleaf) Mental preparation is important prior to running in a barrel race. Sometimes just you and your horse need to go somewhere quiet to prepare mentally.

Imaginary runs also can help. If you're fortunate enough to have six or eight horses, you can work all of them. But most riders only have one horse to practice with. If that's the case, you must work "mentally" by focusing on an imaginary barrel pattern with an imaginary horse under you. The horse should first be the horse you compete on and later it can be horses that require different types of communication during the running of the barrel pattern. You should ride the imaginary horse properly in your mind to practice perfectly. Perfect practice, not just any type of practice makes for improvement!

But it's important to remember that practice, even in your mind, does not make perfect. Perfect practice makes perfect. So when you are practicing, make sure the conditions are as near as possible to the same as they will be when you run barrels. This includes the horse. But you must practice on the type of horse you will be running, whether it's one that turns on all fours or one that turns on the rear end.

Each horse has his own style of running and his own mental condition. It doesn't matter if the same person trained all of them, their styles and minds can still be different. If correct form is not maintained when you practice, speed is your enemy, just like it is when you are training. Instead of "all-out" runs (running as fast as you can), make both you and your horse think by practicing with "tuning" runs (runs made at a slower speed, while refining your horse's maneuvers). We do this mainly with horses that have been away from competition for a few weeks. Don't go all out and be aggressive on a tuning run; run with reserve and make a thoughtful run.

When you come out of the arena after practicing, remember everything that went on. If you blank out anywhere in the arena, you've let your nerves get to you. The nerve factor is the lack of control over your emotions. Your emotions must be kept level, not too high and not too low, for your mind to react quickly and properly. If you let your nerves get to you, you need more practice, mentally and physically.

It's human nature for some barrel racers to have a weakness or problem at one of the three barrels. The rider may just think that a certain barrel is his or her bad (problem) barrel because of something that happened in the past. If you do have a problem with a barrel, think about it carefully and make sure you have the problem worked out mentally before attacking and correcting the problem. For example, the horse that doesn't set or slow down properly for a barrel, must be stopped at the proper position or slowed down until he learns to rate at the proper point by himself. Or the horse that doesn't allow for a proper pocket should be trained to emphasize the pocket until his memory retains the thought of the pocket while he is going around the barrel.

When a horse works correctly two or three times, which doesn't mean perfect, but a little better than he did the last time you worked, don't keep grinding on him at home in the practice pen. Instead, trail ride, work livestock or rope. Do anything to not stress the horse. Don't keep asking for more and more. You can quit for that day and not even work him the next day. Ride your horse and use him, but don't make it stressful or put him under a lot of pressure. Never make practice tedious and boring or let the horse become sore, mentally or physically.

When your horse does something good, reward him. That doesn't necessarily mean to pet him or feed him. It means quitting the training session, unsaddling him, washing him off and cooling him out which may include tying him in the shade. A horse understands praise. He understands when you're telling him that he's done well, but he also understands when you're telling him that he has been bad.

If a rider is just learning about barrel racing, he'll sometimes have to use a horse harder than he should in order to learn. The inexperienced rider might ruin the horse if he works him too much. So when you are working a horse, remember that you want quality work instead of quantity whenever possible. When an older horse is working and running well, don't work him on the barrel pattern unless you must lope big, easy circles. However, if you haven't made a run in several weeks, go somewhere and make a half or 60-percent run. Don't go all out and be aggressive; run with reserve and make a thoughtful run. Don't keep working a horse just because you need the practice.

Before Martha won the high-paying Sweepstakes during The Texas Barrel Race in the John Justin arena (Will Rogers Complex, Fort Worth), she mentally checked out the angles to the barrels. Prior to the race, Martha walked her horse into the arena to test ground depth and arena conditions. (Kenneth Springer Photo)

We'll work or tune a horse from three to six times a week. Six times a week is for a hardheaded, aggressive horse, while three times per week is for a horse that is willing to learn and listens to the rider. We'll work a good horse every other day, but still ride him and do some drills away from the barrels on the other days. This procedure applies to a horse in training.

Walking the barrel pattern is too slow for a horse to learn any advanced barrel education. Running is too fast to put it all together. Medium speed and medium work are ideal to tune a horse. You can work a horse 15 to 20 minutes, slowly and easily, and he'll get softer and softer, as long as there's no pain involved, mentally or physically.

Anytime you have a horse that just quits working, the first thing you look for is pain. But the second thing to look for is something that happened that may have confused your horse. If a horse quits working, it's hard to tune on him if you don't know why he quit working. A horse doesn't just quit working without a physical or mental reason. If you were at a rodeo, try to remember if someone in the stands threw something, or whether your horse may have slipped or was distracted by some noise. Be sure to be objective as to why your horse quit working. There's one important thing to remember: if a horse quits working gradually, the problem is usually caused by the rider.

Barrel racing is a team effort. It's a horse and a rider. Both must do the right thing at the right time to win. Sometimes the rider is better than the horse, so the horse needs tuning. But sometimes the horse is better than the rider, so the rider needs more education and practice. Both the horse and rider need to put themselves in many different types of situations to prepare for the unexpected when they are competing for the money.

A horse can't imagine what may happen at a barrel race, but you can. You can use your mentality to be prepared for the unexpected at a barrel race. You can do this by checking out the arena when you first arrive and walk into it. You should first look at the angles, where the stakes are, where the fences are located, where the gate and alley are, and where the solid fence or alley is that you've got to come back to.

Mentally prepare yourself, right then, so that you are aware of all the visual things around you and decide what you are going to do. Never come into an alley cold (never having been there before). You must do your homework in advance. After you have looked at the arena that you will be running in, go over your run again in your mind; this time mentally in that arena.

Ten minutes before you are scheduled to run, walk your horse. You should have already loped and warmed him up and he should be walking and getting his oxygen. During that time, don't spend time visiting. Think of a perfect run. Your winning run depends on your mental preparation along with how your horse feels under you. As you get more experienced, you will be able to tell when things are really right and when things are just so-so.

(opposite) An announcer's stand above the alley can unnerve a green horse. It is important to walk your horse under the announcer's stand before running him in a barrel race.

If you're not ready, take your time going into the arena. Try to get that right "feel" before you get to the gate. On young horses, sometimes it's like grasping at straws, because they're usually not focused or listening to you. A young horse will be looking around unless you have him focused.

If your horse is looking around, try to draw his focus back to you. In most cases, putting your horse into the bridle or collecting him up will bring his focus back to you. You can collect your horse by using pressure from both calves of your legs, both hands, your seat and your feet. Collect your horse to where you want him to be when you get to the barrel. Remember to be very soft and light with your hands. When you keep your horse mentally focused and listening to you, the wheels in your mind are also turning better. The more tense and more electric you get, the slower your mind reacts and the slower your body reacts. Soft is fast, hard is slow, both physically and mentally.

If you have a horse that isn't well-educated to turning the first barrel, you can't expect to compete with a person who has a horse that has three good barrels. In a case like that, you place where you're supposed to place ... behind them. If there are three good horses there, you place fourth. There are a couple of reasons for this. If you try to win first and your horse is weak in education and training, nine out of 10 times, it will be a total wreck and you won't win anything. You'll win a lot more by placing fourth and having a smooth run.

But the most important reason is that in order to advance, you must keep your horse turning a good, solid first barrel, making no mistakes. It doesn't matter how hyper your horse is. If he will work it and go on, it will eventually be equally as good as the two good barrels. It may not happen in two weeks, or two months, but eventually it will happen on 90 percent of the horses. When you do make your run, remember, you're not competing with other people. You're competing with your horse with those particular ground conditions and on that particular pattern, always keeping in mind your horse's capabilities. Your challenge should be to get the best run you can.

One of the hardest things for a novice rider to do is to get a horse to think, to use his brain. In order for a horse to be educated, the rider must be just as educated. Everyone can run barrels. However, the main key is how much you want to put forth to excel.

You've got to experience a number of competitive situations to really advance in your mental preparation. You can't do this totally in a classroom environment or by working a horse at your home. You've got to go away from home and be under stressful conditions for your horse to mature. Also, to stay sharp, you and your horse have got to periodically be under pressure and have some competitive runs.

If you want to correct a problem that you have and think that you may be creating, watch yourself on an instant replay camera. Carefully look at what you are doing. You will miss things if you don't watch closely. For perfection, study what you're looking at. Always analyze your problem in your mind and go over everything you can remember seeing. If you do this, you will learn something every day. We have found there are no two horses that are identical and we learn something from every horse we get on.

MENTALLY COOLING OUT YOUR HORSE

One of the things that is probably the most overlooked, is after you have run, you need to cool your horse down, not only physically, but mentally. You horse can be cool physically, but you still need to cool him out mentally while you're on his back. You need to have your horse quiet and level before you quit.

Leading a horse does not cool him out mentally as well as riding him cool. A rider shouldn't just get on a horse before competition and then always get off the horse after competition. If you do that, your horse knows that when you're on him, it's stress time, and when you're off him, leading him, it's relaxation time. It's better to ride the horse to a cool physical and mental level after competition to keep his mind soft and relaxed.

"WE LIKE TO PRACTICE BY SIMULATING THE RUN AS MUCH AS POSSIBLE."

12

Hauling Your Horse

TRAILERING

When you haul a horse, how you drive and handle him really affects how competitive that horse will be when he gets there. If your horse isn't used to being in a trailer, and you drive for 11 hours and never unload him, he's not going to be very good mentally or physically, when you get there. If a horse isn't ready to haul and is not physically fit, unload him every two and one-half to three hours and let him move around. If you're in a jam on time and the horse is "road ready," you can go four or four and one-half hours before unloading, but not any longer.

A lot of people wrap a horse's legs when they haul them but we don't wrap legs for any reason unless the horse has a leg problem. (For further information on wrapping legs, see Chapter 15.) For hauling, however, we do suggest shipping boots so a horse doesn't step on himself.

We sometimes feed our horses when we haul them. Even though we have heard of some horses choking in a trailer when they are eating, we feel you can take precautionary measures to prevent such a thing from happening. You can feed your horse in the trailer at home several times to teach him to relax in the trailer while eating. Also, if you drive safely and cautiously at all times when pulling a trailer with a horse in it, it will help make the horse feel safe and comfortable in the trailer. That way, he should be under no stress while he is eating in the trailer.

You can use a hay bag in a trailer and hang it in such a way that the horse can't get his legs or head hung in it. The conventional rope-style hay bag is very dangerous because when it's full, it stays up, but when it's empty, it hangs down and a horse can get hung up in it. Also, a horse can chew on it and swallow pieces. We prefer hay

Trailer floors need to be adequately padded with shavings on top to cut down on the smell of ammonia.

A trailer should be high enough so the horse can lift his head without hitting the roof.

(overleaf) The ground is a very important factor when running barrels. Always check the depth of the ground before you run. If it's deep, you'll have to run your horse further into the pocket before you rate him. If you check him too early, he will have a harder time getting through the deep ground. (Kenneth Springer Photo)

bags that are solid with a big hole in the center. The horses don't waste as much hay out of them either.

The trailer floor needs to be adequately padded. We use big, heavy stall mats and we put shavings on top. Shavings keep the ammonia smell down by absorbing the urine, and they help keep urine acids from contacting the horse's feet. Also, horses will urinate on shavings quicker than they will on just a rubber mat, which causes urine to splatter. A lot of horses don't like to be splattered and, therefore, won't urinate in a trailer. For obvious reasons, it's important that a horse urinates when he needs to. However, shavings can be dangerous if sparks or cigarette butts fly into the trailer. Always keep a fire extinguisher with you, not only for the horse and the trailer, but for your tow vehicle as well.

Trailers need to have adequate room. A horse needs both width and length in a trailer to be comfortable. He also needs room between his chest and the end of his nose by having an extra long feed manger area, so that if you stop, he won't hit his head. The trailer should also have enough height that the horse can pick his head up if he needs to and not hit the roof. A trailer also needs insulation, especially in the roof and certainly in hot country like Texas or Arizona. Otherwise, it's like putting a horse in a sauna. Color is also important. Black trailers absorb heat while a white or light-colored trailer is the best for reflecting heat off the trailer.

When you get to where you're going, it's important for your horse to have shade. A regular plastic tarp won't work. That's like putting a heat absorbent shield above the horse. Shade tarps that come from nurseries and that let a certain amount of sunlight and ventilation through are the best. A horse has to get used to any covering, because when a tarp is flopping in the breeze, the horse is going to be frightened, tear something up and leave town if he is not used to a cover. Tarps used for shade are easy to hang. You can go across the top of your trailer with a rope, park next to the fence and tie the rope to it. Metal posts can be carried and driven into the ground for shade support.

Anytime you're going to stay at one place for a few days, your horse needs a place he can lie down. That old theory that a horse can stand up and sleep is baloney. A horse can stand up and relax some, but when he really sleeps, he needs to lie down just like we do. A horse doesn't sleep eight hours, he may sleep only 15 or 20 minutes at a time but when he sleeps good, you can walk up on him and even say something and sometimes he doesn't even know you're there. The horse can be in a deep, sound sleep.

ARENAS

Before going to a barrel race, we like to practice by simulating the run as much as possible. We'll even set the size of the barrel pattern at home the same size of the barrel pattern where we are going to compete. For instance, if we are going to a small, indoor arena, we like to run our horse somewhere away from home, in a small, indoor arena.

We may even lease an arena somewhere that simulates the same conditions where we are planning to compete. There we put together a competitive run that is as close to a competitive situation as we can get. If we have a green or young horse, or if we are going to compete under extraordinary conditions, we try to get to the rodeo or futurity early and, if possible, work the horse in that particular arena.

The Astrodome arena in Houston is a very unique arena. There is no way one can simulate it anywhere else. Even when the best, most-seasoned horses get there, they need a practice run before the event. Barrel horses that compete at Houston need to get tuned up in that arena if they are going to run competitively during the rodeo.

What makes Houston different is where the walls are, the acoustics, and where the crowd is located (above you at the starting gate). A horse's depth perception is based on the location of the fences and the height of the roof of an indoor arena. The roof at Houston is so high that a horse doesn't relate to it as being a roof. A horse seems to get his balance off the distance between the fence and the barrel. The depth perception is so different at Houston that a horse has a hard time getting that balance.

If your first barrel at Houston is your right barrel, it is moderately close to the fence. Your second barrel is out in the open with no fence anywhere in sight. That will get a horse a little disoriented if he hasn't been tuned or worked in the Astrodome previous to the time you run. The third barrel has a fence, which in reality, is a line of bucking chutes. All three barrels have totally different backgrounds, so you must have a horse that is accustomed to that situation. Even seasoned veterans need to have a training session in the arena prior to competition.

On the other hand, Mesquite, Texas, is a really small arena, with hard-packed ground. You want your horse to really have set and turn on his mind because of the noisy, distracting conditions. However, if your horse is the type that sets and turns too much, you must have him a little more free because the people are right in your face and the wall is right against you. Also, you should probably practice on some ground that is a little harder if you're planning on running at Mesquite, so your horse can learn to get his feet under him and stand up on hard ground.

Vernon, Texas, has a large rodeo pen with deep ground and it's a long way between barrels. If you have a right-handed horse (one that goes to the right barrel first), the second barrel is right up against the chutes. There is usually stock in those chutes, so you need to get a horse seasoned for surprises like that. He needs to learn to look for the barrels. So if we're going to Vernon, when we're practicing at home, we'll set up a really large pattern.

If we are going to Mesquite on Friday, we work our horse accordingly Friday morning. If we're going to Vernon on Saturday night, we set up a big pattern at home and work it. However, if we leave Mesquite and go directly to Vernon, we would try to be there at daylight and get the horse in that arena, especially if our horse is not a seasoned veteran that can adjust to going from one size of arena to another. However, a horse that is an older, seasoned veteran, with a good rider on him, can adjust during his run. This horse should not need to be worked ahead of time.

HAULING TIPS TO HELP YOU AND YOUR HORSE

➤ When you're at an arena, don't park close to where you come into the arena. Horses tend to "buddy up," and they know where your trailer is parked. If the only available parking is close to the gate, go in and tie the horse on the blind side (where he can't see the arena gate), as far to the back as you can.

➤ Don't ride around the arena to warm up and then sit in the corner and visit with your friends. That lets the horse think about going down to that corner. If you want to visit, keep in motion or go out of the arena and visit.

➤ When you go to a rodeo, get in the alley prior to the start of the rodeo, so you can see where that first barrel is. You should study your angles to the barrels and look at what's behind them. Don't come through the gate at a dead run, with both you and your horse guessing where the first barrel is.

➤ Check the depth of the ground for how hard it is before you run into the arena. If the ground is hard, you should rate your horse and get him to stand up (not lean into the barrel), where he can use himself the quickest and safest. If the ground is deep, you'll have to go farther in toward the pocket before you rate your horse. If you check him early, it won't be easy for him to get through the deep ground. He'll hang up in the deep ground and have trouble getting around and away from the turn.

➤ Always pay attention to where you're going to exit the arena and what's out back. If it's concrete, you'll know that you're going to have to get shut down before you get to the hard surface area. Obviously, you must get across the electric eye at the finish line before you stop. But don't watch the eye while you're running home. Before you leave the gate, figure out where you want to slow down after crossing the electric eye.

➤ The most dangerous thing about barrel racing is running back to a closed gate. If you do this, do it like turning a barrel. Run toward the fence, rate and always turn the same way. It doesn't matter which way you turn, but train your horse to always turn the same direction. If you train out in the open field like we do, get four panels, set them up, run back to them, rate and turn the same way every time. That will help train your horse to run into a closed gate or fence. The horse that runs to the fence and guesses which way to turn is dangerous to himself and you. If you're at a rodeo and the gate is to the left, and you usually turn right, the producer will probably scream at you, but turn right anyway, then get out of the arena.

➤ If you come into an arena and they close the gate behind you, trot down the fence, get your momentum up and head for the first barrel. Or, if they let you, make a circle the same direction as the first barrel so you are on

If you are running in a big rodeo, get in the alley prior to the start of the rodeo so you can see where the first barrel is. Study your angles. Martha won the big rodeo at the San Antonio Livestock Show and Rodeo. (David Jennings Photography)

the correct lead. If they don't let you circle, turn and head for the first barrel. Letting a horse stand still forms a hot spot (nervous place) for the horse.

➤ Never walk down the fence or come through the alley at a walk when you're going to run the pattern. Don't walk up an alley, stop the horse and then start running. This forms a hot spot for the horse. We like to start far enough back to allow room to trot and then lope up the alley with progressive speed, where we can keep the horse's mind level. Ride soft, don't get wound up and tense unless you have a real deadheaded horse. If you ride soft, it will keep your horse levelheaded and focused.

TIMERS VERSUS FLAGGING

We still go to rodeos or jackpots that use humans to stop and start the time on the finish line. We don't mean that they control the time in the clock, but they use a flag when the horse crosses an imaginary line to signal the secretary to manually start and stop the clock or timer. An electric eye, with an electronic beam that is broken when your horse runs through it on his way to the first barrel and breaks it again to stop the clock when he leaves the arena, is a much more precise and advanced method. It does away with human error, human reflex action and a human's short attention span.

It is important that the jackpot, rodeo or futurity that you go to uses an electric eye, rather than a flag, to keep time. With the quality of horses that are running today, one-hundredth of a second can be the difference between first and second place and there is no way that a human reflex can record time down to one-hundredth of a second. The only way this can be done is with an electric eye.

GROUND CONDITIONS

Anywhere you compete, the ground changes, even if it is a futurity where the show managers definitely care about keeping the ground consistent and in good shape. When there are a lot of runs made, the ground conditions change. It's almost impossible for the ground to remain the same through a large number of runs. The ground can even change five or six times during the day.

The moisture content of the ground naturally varies according to the wind and the outside weather conditions. The depth of the ground varies according to how it is plowed and what was done in the arena the week before. For example, if there was a tractor pull, there's a definite disadvantage to having a barrel race in the same arena afterward, because the ground can be extremely hard and difficult to work up.

Perfect ground for a barrel race should have a clay base topped with a loamy substance mixed with sand. Martha takes most of her young horses to practice in a big outdoor arena in Stephenville prior to competing.

The best preparation for the variation of arena conditions is to prepare our horse at home. You must fix your ground at home so that sometimes it's dry and pushy (the ground moves easily and does not provide safe footing for the horse), and other times it really holds your horse. Haul your horse away from home — to deep ground, pushy ground, dry ground and perfect ground. That is why seasoning is so important for a young horse. They must be exposed to all of these conditions several times. When they run into an arena, they need to learn to "feel" the ground themselves.

Most arenas present some variation in conditions. One pattern might have the first barrel on your right, immediately when you come in the arena. Another might have the first barrel way down the pen where you have to go a long way before you turn it.

In the perfect arena, the ground would be prepared the same way it's done at some of the top racetracks, which means that it has a cushion or soft surface on top and adequate firmness underneath to provide both "soft and hold" conditions to run or work on — soft to provide a shock absorber and hold to prevent the horse from slipping.

To prepare this perfect ground, there should be a clay base down deep in the ground to hold the moisture after it's put in the soil. Above the clay, there would be a loamy substance that is mixed in with sand. Loose sand is bad to run in and turn on because it doesn't hold a horse in a turn and in the straightaway; it tends to pull a horse down and really stress him, especially if it's too deep.

You cannot have hard pan (a hard layer of dirt) a few inches below the surface. That means if you plow six inches down, it's hard as asphalt. If it's hard underneath, with a real loose, powder-type sand on top, that won't hold a horse. The body of the sand must have a slight amount of loam mixed in it, but not the packing kind. Then when moisture is added, it will form a good substance that will hold and support the horse.

Some locations have a lot of hard ground. We'd rather have ground too deep than too hard because obviously a horse uses himself harder and is more susceptible to injury on extremely hard ground. It is also harder for him to stand up and in the long run the horse's welfare is what we have in mind.

Also, at home, we don't keep the ground where we run barrels level at all times. That's because we sometimes compete in arenas that aren't level. We feel that by keeping our arena unlevel, the horse learns to feel the ground and he's careful where he places his feet.

*"EACH PROBLEM USUALLY
HAS A SOLUTION."*

13

Solutions for Common Problems

Because of the speed used in barrel racing, there can be a lot of problems, though each problem usually has a solution. Some problems are caused by the way a horse is bred, such as too hot or too much Thoroughbred in him. Other problems are caused by the way he is built and others are man-made, as a barrel horse can easily be spoiled if the rider doesn't know what he's doing. Or maybe the horse just isn't completely trained.

A HYPER HORSE

If a horse is hyper and nervous, we're liable to do what some people would consider "unorthodox things" to make him relax, because we want him to retain the thought of what he just did. We might even get off and pet him at the spot where he becomes tense and nervous.

To get a horse to relax and calm down, sometimes it takes fatigue. In that case, we would lope a horse in a big circle until he relaxes. Another way to relax a horse is to step off him, uncinch him and let him stand and relax. Occasionally, we get a horse that becomes really hyper just in one spot. If that is the case, at that particular spot, we make him slow lope until he relaxes.

SPOILED HORSES

Retraining a spoiled horse is always harder than training a horse correctly from the beginning, because later on when you put a little bit of pressure on that horse, he will probably go back to his old habits. To a certain degree, bad habits will always be in a spoiled horse's mind.

When we get a spoiled horse, we go back to basics. Obviously, when we have a spoiled horse, we have to change some of his habits. Therefore, it requires a certain amount

of brainwashing as far as the horse is concerned. That means we need to use some "shock treatment" to get the horse to think and accept new and different education. First we remove the bad habits by making him listen to us physically. We use fatigue as a tool for punishment and to get a horse to relax. Then we go back and show him how we want him to change mentally through repetition.

GATE OR ALLEY-SOUR HORSE

A lot of horses that are difficult to start in an alley or gate are called "gate-sour." Youth and novice riders often cause this bad habit because they do not do enough slow work. The less knowledgeable rider seems to put a lot of tension into his or her ride instead of riding soft, quiet and easy, even with speed.

If a horse does not want to come into the arena, lope him into the arena and head in the opposite direction of the first barrel. Lope around the perimeter of the arena and out the gate. Make a large circle outside the arena, still at a lope, and then go back into the arena and do the same thing. Do this at home until the horse is totally quiet. Then, when you have it corrected at home, repeat the process at another arena. If you don't take your horse away from home for this exercise, he will likely go back to the same old habit of not wanting to come into the arena. The problem has to be corrected first at home and later away from home in a competitive environment.

The first time you leave home, you'll probably have to go someplace where no one else is around, because you won't have time during exhibitions at a barrel race to properly educate your horse through slow, quiet work. Then, you must educate him again under competitive conditions. A horse with a problem, once it is corrected, must stay corrected from day to day. Each day he stays corrected it is easier for him to remain corrected for a longer period of time. Although, you must be aware that the thought still exists in the horse's mind and you must always restrain it. Try to never put the horse in a position or situation that presents the opportunity for him to regain the old habit.

Also, you must ride consistently. Any time that you leave home, be sure to ride the same way that you do at home. That requires thought. Some people change the way they ride when they leave home. That puts pressure on their horses because they are suddenly riding differently.

RUNNING PAST THE BARREL

Some horses run past the first and second barrels, creating problems for their riders. Too much uncontrolled speed in the approach to a barrel can be the cause of a horse running by a barrel or getting out of position to turn a barrel. When a horse runs by a barrel, it's usually a rate problem, compounded by being in the wrong position. Correcting the rate problem can easily be taught to the person and the horse. However, teaching proper position on the turn is harder. Both of these solutions were covered in Chapters 7 and 8.

(overleaf) Many common problems can be overcome. Although Martha was riding a hyper horse, she won the 1998 Elite Barrel Futurity in Glen Rose, Texas. "I sent her to a horse trainer a week before the Futurity to just ride her and calm her down," Martha explained. "And it worked. She didn't need anymore training — just calm riding." (Kenneth Springer Photo)

When correcting a horse that has a "hot spot" (a place where he is nervous and highly stressed, thereby running by a certain barrel), take the horse to the point where he's the most stressed. Start loping big circles right there until he's tired. When he's fatigued in that position, stop and let him rest there. You can even get off, brush him, clean his feet and get him relaxed in that position. That's disciplining him through fatigue.

Much of our education and discipline is done through either repetition or fatigue. With a young horse, we educate him to where he clearly understands through repetition. But the problem horse is trained with repetition and fatigue as a discipline.

TURNING BEFORE THE SECOND BARREL

If a horse starts to turn before he gets to the second barrel, or runs wide on the completion, there could be several things involved. First, look for some pain coming from somewhere, because pain is usually the first indication of an execution problem. Something could be pinching the horse inside of his mouth. This can happen when the rider hasn't checked the bit's mouthpiece to see if it has a sharp edge or he hasn't put bit guards on the bit.

Next, check the saddle. A saddle with a broken tree is a major problem. A horse relates to that pain quicker and quits working faster than from any other problem. It's also possible that you could be doing something totally out of the ordinary. Maybe you're actually cuing the horse to turn and don't realize it.

One of the most obvious (and least thought about) reasons is riding to a place where the horse's vision is affected and things are different to him. This could involve depth perception or a sign on the fence. Maybe the barrel is the same color as the fence. At a rodeo arena that used to be in Grand Prairie, Texas, we saw more horses duck off going to the second barrel than any other arena we've been in. It's probably the way the fence was set up; the background behind the fence (like the bleachers), the slotted wall, the lighting from the sun or the arena lights.

Last, but not least, your problem could be that when you trained the horse at home, you set up the barrels with a shorter distance between them. Believe it or not, horses count steps. If you leave your barrel pattern the same size, your horse expects to turn after so many steps.

COMING OUT OF THE THIRD BARREL WIDE

When a horse comes out of the third barrel wide, it is usually a position problem created by either the horse or the rider. Many things may have happened. The first may be that you started your turn too early, forcing the horse to have to take a step off to the side in order to finish his turn. To correct this, remember that not only the horse's nose, but also the horse's spinal column, must be facing the finish line before he starts to run. You need to make sure that you are in the proper position going into the turn.

Then, if the horse tries to run before he has completed his turn, totally over-turn the third barrel with a little speed. Then do the loop (see Chapter 9) at the third barrel, also with some speed. You can also go around the third barrel two or three times with moderate speed to teach him to totally finish that turn.

Your horse could also come out of the third barrel wide when he has not rated going into the turn and he is trying to make the turn going too fast. To correct this, make sure you have set and rated the horse and that you have picked his shoulder up as you go into your turn. Make sure he is not pushing down on you or on his front end because that will cause you to lose control of his position and his body. The problem can also be caused by you. Once you have rated your horse going into the barrel, you must be aware of the point on the ground where you will ask for the turn. Don't ask for the turn too early.

In the heat of the turn, some riders want to "pull" that turn out of the horse. When you pull the horse, he will want to pull back against you, which makes him pull out of the turn. Instead, you should use a give-and-take on the reins, which asks that horse to come around, instead of telling him. That way, you will get a lighter and better response.

HITTING BARRELS

Hitting a barrel while coming into it is usually due to lack of control. Some horses, as they approach the barrel, drop their shoulder and turn their head to the outside. For some riders, their first reaction is to take that outside rein and try to hold the horse off the barrel. But that creates a bigger problem. When you take the horse's head out, that pushes the shoulder in and the shoulder is going to go right into the barrel, tipping it over.

The correction for this is getting the horse's rear end underneath him, allowing him to stay mobile in the front end. As you approach the barrel, pick the horse's shoulder up just slightly (with the reins) and ask for a slight amount of flex in the shoulder and neck area. That way, he is prepared to make that turn when he gets there. You must have light hands. Generally, the more you pull on a horse to keep him from doing something, the more he will pull back. Learn how to communicate with your hands and educate the horse to react and think properly.

To help a horse which tends to hit barrels going into them, lope big circles around the barrels. That teaches a horse to listen to your hands and go where you want him to go. You, not the horse, must gain control and call the shots. If you are hitting a barrel coming out of it, it is usually a matter of being too close to the barrel. Maybe you've made that perfect turn, but you've made it just a little too tight.

However, if you hit barrels consistently, then it's more than likely a position problem. You're not keeping enough pocket around the barrel (the area between the horse and the barrel) in order for him to get around it cleanly. It may

(opposite) If your horse is hitting barrels, more than likely you didn't move the horse's shoulder out prior to going into the barrel. You can pick up the shoulder with the inside rein and ask for a slight amount of flex in the shoulder and neck area.

Martha tries to neck rein her horse to keep him from hitting the barrel. Most of the time, neck reining a horse off the barrel will just compound the problem. The more you pull on a horse to keep him from doing something, the more he pulls back.

To keep a horse from dropping his front end and hitting the barrel, ride your horse into the bridle, which collects him, then elevate your hand a little to put him on his hind end.

be because you're picking your hand up at the point of the turn or maybe you're asking a little too hard.

Also, if you're hitting barrels, either going into or coming out of them, you must overemphasize the opposite. When you go in, at the very point where the horse hits the barrel, move the horse out about three feet at that point into a large circle. Then, when you come back in, long trot to work the horse close, then move him out about three feet right at that point. The reason you move out is not to teach him to move out at that point, but to teach him to hold his area when you come to that spot.

Obviously, this calls for judgment on your part because you can move the horse out for such a long period of time that you make him think he's supposed to move out and stay out, rather than stay off the barrel. The horse should maintain the proper amount of forward motion and adjust this with the correct amount of rate. Together, these equal a perfect run. You must decide why the horse is hitting the barrel. Sometimes he might be locking his rear end under him and rotating over his hocks instead of running around the barrel. Or he may have disengaged his rear end too much, which causes the rear end to swing too far and brings the horse around the back side of the barrel too soon.

A HORSE THAT DROPS ON HIS FRONT END

If you have a horse that drops on his front end, he usually shoves his face out, his shoulders are really flat and he raises his rear end and wants to get tight when he comes into the barrel. This is not a good position and it is very hard to ride and control the turn. To keep your horse off of his front end and put him on his hind end, elevate your hands a little. You don't want a horse that runs through the bridle (having no regard for the bit). Ride a horse "into the bridle," which collects him. Elevating the hands slightly should level the horse.

We want a horse to reach, running as close to the ground as possible between the barrels, but we also want him to be collected when he's turning them. If he's leaning into the barrel, he can't turn the barrel as sharply. Also, he can't stand up on bad ground, and he can't leave a barrel as quickly. When he's collected, he can stand up on bad ground and leave the barrel quickly.

HORSES THAT SET UP TOO MUCH

If you're asking a horse to run and he has a place where he rates or slows down too much, you need to help him free up mentally. If you have a horse that sets too much going into the third barrel, a stride and a half before the barrel, use an over-and-under or a quirt on him once to free him up. After that, go right back and help him with your hands if he needs it. Educate the horse to free up with large circles.

For a novice rider, we would not suggest using the over-and-under on your horse going to the first barrel. But if you do use it, start in the alley while you are moving forward. As soon as you come around the first barrel and get into position for the second barrel, use your over-and-under. But most novice riders can't effectively use it between the barrels and still get prepared to turn. Also, using the over-and-under that late doesn't let the horse's mind settle so he can get ready to turn.

If there's a certain place that the horse wants to set when he's not supposed to, use your over-and-under there. But normally, you should use it only as you leave each barrel. You need to reeducate your horse to change the problem. The over-and-under is only a temporary solution. When you leave the third barrel, don't hit your horse as fast as you can. Hit him and then let him make three or four strides before you hit him again. The hit, however, should be more like a tap. When a person's adrenaline is flowing, he or she can unintentionally hit a horse hard enough to knock the run out of him, rather than hitting him softly enough to ask him to run.

Some horses make up a lot of time in the turn. Other horses make up for it in between the barrels. Obviously, the ideal horse can do both, but not many of those exist. If a horse has a weakness in his turns, then he has to have some speed between them to be highly competitive. If a horse has some weakness in his speed, he can't have a flaw in his turns if he is going to win. Assess what talents your horse possesses, then minimize his weaknesses and capitalize on his strong points. The same rules apply to the rider.

"A HORSE IS AN ATHLETE AND MUST BE MAINTAINED."

14

Health Care and Nutrition

A horse is an athlete and must be maintained. You need to use the services of a good veterinarian. However, your veterinarian may feel there are some things that you are able to do yourself to keep your horse in good health. Before trying any of the following, consult with your veterinarian.

INOCULATIONS

Inoculations are imperative. Your horse is exposed to germs and disease at all times. If a bird lands on your place or a mosquito ever comes in, your horse has more than likely been exposed. Some shots that you can give your horse are designed to prevent tetanus, rhinopneumonitis (rhino), influenza (flu), streptococcus equi infection (strangles), rabies and encephalomyelitis (sleeping sickness).

Giving a shot can be dangerous if you don't know what you are doing. A person can be pawed, kicked or run over by a horse, if he has not been taught by a veterinarian how and where to give a shot. If you don't know how to give a shot, don't. Take your horse to a veterinarian.

If you do give the shot yourself, and your horse stands in one place too long after he has received a shot, he may experience some soreness. Therefore, immediately after a horse receives inoculations, he needs to be exercised. Lightly exercise him to get the vaccine to spread out and be absorbed. A horse can get sore from a reaction to the inoculation which has been given to him to help his immune system. Exercise helps the body's circulation take care of distributing the contents of the shot. However, use discretion. Don't take a horse out and overuse him after he has had a shot. Give him only light exercise.

Tetanus: A tetanus shot protects against Clostridium Botulinium, a type of bacteria that invades open wounds or punctures, and that can cause a potentially life threatening disease. Every horse should be vaccinated against tetanus. A tetanus shot should be given to a horse annually and re-administered if he has suffered a wound.

Encephalomyelitis (Sleeping Sickness): A sleeping sickness shot protects against viruses that cause life threatening neurological disease. There are three forms: Eastern (EEE), Western (WEE) and Venezuelan (VEE). There is a four-in-one shot available that protects against all three forms of sleeping sickness, plus it has a tetanus toxoid included. When the first four-in-one shot is administered in a two milligram dose, then in four to six weeks use a different injection site another two milligrams are given. After that, only a single annual booster is required.

Influenza: A flu shot protects against one of the most common viral causes of respiratory disease. Although influenza is rarely fatal, it can take your horse out of action for several weeks or more. Therefore, it is a necessary shot for young horses under three years of age and show horses. It is also imperative for horses that have been exposed to many different environments and other horses, such as at boarding stables, pregnant mares and horses with existing respiratory problems. High-risk horses should be vaccinated every three months, and pregnant mares one month prior to foaling. For "stay-at-home" horses, a biannual, or even an annual vaccination is acceptable.

Rhinopneumonitis: A rhinopneumonitis shot protects against one of the most common forms of respiratory disease. Even though all horses should receive rhino shots, it is of particular importance that pregnant mares receive a rhino vaccine, the EHV type 1 that protects against abortion. Pregnant mares should receive this shot at months three, five, seven and nine and EHV type 4 one month prior to foaling.

Rabies: A rabies shot protects against the neurological disease hydrophobia, caused by a virus that's generally transmitted by a wild animal bite. If your horse lives outdoors in a high-risk area, rabies vaccine should be given annually.

Streptococcus equi infection (Strangles): A strangles vaccine should be given to healthy horses of all ages as an aid in the prevention of strangles due to Streptococcus equi infection. A one milligram shot should be given intramuscularly, preferably in the hindquarters. The primary immunization consists of three doses given at intervals of three weeks. Foals vaccinated when less than three months of age should receive an additional dose at six months or at the time of weaning. Revaccinate annually and prior to anticipated exposure, using a single one-milligram dose.

SWELLING

When you have a little swelling in a leg, but there isn't any injury, the best way we've found to pull the swelling out within eight to 12 hours is with white vinegar, a brown

(overleaf) Learn how to wrap a horse's legs from an experienced veteran before trying it yourself.

grocery sack and a wrap. In most cases, this method is quicker than anything else you can use. The brown paper bag helps draw the swelling out of the leg better than any other wrapping. It is extremely hard to wrap without a wrinkle, but when you're done, it accomplishes much more than any other method we have found.

Wet the horse's leg with white vinegar and rub it into the leg before wrapping it. Then wet the brown grocery sack with white vinegar and wrap it around the horse's leg. Cover the paper sack with a quilted, cotton leg wrap with a plastic polo wrap for support. Use even pressure all the way up and down and be aware of how to wrap a leg, wrapping from the outside toward the inside, and use consistent pressure from the top to the bottom of the leg.

One caution, however, if you wrap a horse's leg, you must be very sure the wrap is smooth all the way around the back half of the leg. If you get a wrinkle down those tendons, you can bow a horse's tendon. Also, if your wraps aren't even, the same thing can happen. If you wrap loosely part way down the leg and then pull real hard right in the middle, you will put a lot of pressure on a leg and you will be in trouble with a tendon injury.

WRAPPING A HORSE'S LEGS

Never wrap a leg unless you have a problem. If you have a swelling in the leg, or you want to tighten the ligaments, it is all right to wrap. But if you start wrapping a horse's legs all the time, the horse will become dependent on the wraps for support and his body will quit taking care of the tendons and ligaments on its own. It is probably better to just put something, like a liniment, on a leg to tighten it rather than wrapping it.

If you wrap a horse's legs for hauling or shipping, you leave the most damageable part of a horse exposed — the coronary band — as most people don't wrap that far down. We do, however, believe in shipping boots when hauling a horse. They should go from the knee and the hock all the way to the shoe. That protects the whole hoof wall and coronary band.

Shipping boots need to be cleaned regularly because they get dirty quickly. The shipping boot itself should be lined with a slick material. It does not need a fleece lining that holds moisture, creates friction or picks up what the horse walks over, such as shavings or burrs. We like a boot that has Velcro® closures. If you use a leg wrap instead of shipping boots, you have to be very careful of the pressure applied. If you don't apply a leg wrap correctly, you can do some severe tendon damage by making one wrap loose and one tight.

About the only time we wrap a horse's legs is if we put the horse in a stall while we are on the road. We'll wrap his legs under these conditions, simply to keep them tight and the inflammation out. But remember, the pressure you use in wrapping should always be pulling the tendon to the inside of the horse's leg — from the outside to the inside.

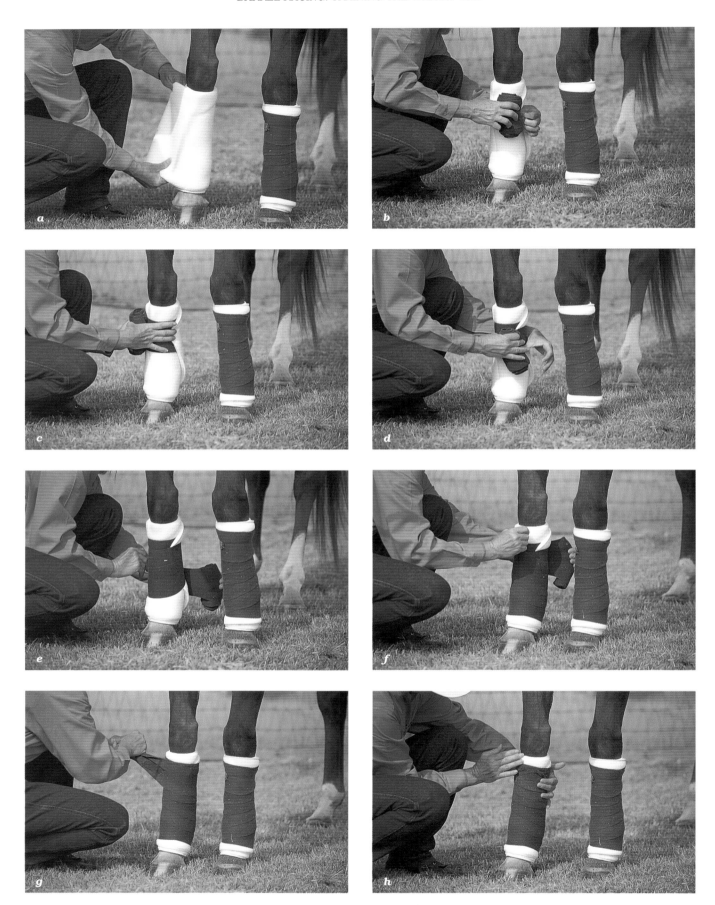

164

For example, when wrapping the right front leg, hold the rolled leg wrap in your right hand. With your left hand, pull about the first six inches of the bandage around the leg, just behind the knee, and then in front of the knee to create a flap. Hold the flap out of the way while you wrap, going toward your horse's head first, starting at the top of the leg and spiraling an inch lower on every overlap. For even tension, hold the bandage straight up and down as you roll, not at an angle.

We use a standing cotton wrap. We don't like fleece because sometimes it will compact and if not kept extra clean and fresh, it doesn't have a consistent wrapping texture. We then use a polo run-down wrap to go around the cotton and hold it in place with even tension all the way up and down (adequate tension to keep the bandage in place but not so tight that we form too much pressure on the tendons, ligaments or restrict the horse's circulation).

Let the material just glide across the tendons at the back. Wrap down the leg and take one turn around the ankle. Smooth the flap you left down the back of the leg to cover the length of the tendons, creating a double thickness there. Continue wrapping the wrap up the leg, keeping the tension even and smoothing the material as you go. If you've never wrapped legs before, it's extremely important that you have an educated horse veteran with experience — one who knows about the proper pressure to apply when bandaging legs — help you when wrapping legs.

DEWORMING

Deworming your horses is mandatory. Work with your veterinarian on an effective program, because deworming should be done only after a fecal test has been run on the horse and it has been determined that parasites are present. This should be done every 60 to 90 days to establish a baseline egg count or "eggs per gram" (EPG). Don't deworm just for the sake of doing it periodically. That's hard on the horse's system, especially his liver.

We don't like to tube-worm anymore. We feel it's too traumatic for a horse. You also need to rotate types of wormers because parasites become resistant to worming drugs if they are used too often. Anytime your chemical deworming is not working and your horse's fecal EPG rises above 200, take a look at your stable management.

To expose parasite larvae in a pasture to damaging sunlight, mow your pasture every week or two, depending on how fast it grows during the summer months. However, to effectively break a parasite's life cycle, you should deworm all of your horses that share the same living space all at the same time. To minimize the ingestion of parasite larvae and eggs, make sure that your horse always has fresh, clean water and a clean feeding area.

HORSES WITH PINK OR LIGHT SKIN

In the summer, you have to be careful with a horse that has pink skin. A horse which has white hair on his face, such as a blaze, strip, bald or apron face or white

(opposite) Wrapping a horse's legs.
a) Prior to wrapping a horse's legs, put a quilting or foam padding on first. The pad should finish in line with the groove on the other side of the leg so that a double thickness covers the back of the leg where the tendons are located.
b) Start wrapping the elastic bandage in the middle of the leg.
c) Work down the leg in an even spiral.
d) Keep a firm tension on the elastic bandage, but don't stretch it out all the way.
e) Wrap the bandage down to the ankle, making two turns around the ankle to secure the bandage. Then start up the leg.
f) If you have bandage remaining, wrap around the top some more; don't work back down again as it could jeopardize the bandage's security.
g) The pressure you use in wrapping should always be pulling the tendon to the inside of the horse's leg — from the outside to the inside.
h) If the bandage does not have Velcro®, you can secure it at the top with two pins crisscrossed on the outside of the leg. If there is Velcro® and it is the least bit fuzzy, add a band of tape or a pin.

muzzle and nose, has pink, unpigmented skin underlying the white hair. That skin is very sensitive to the sun and will blister, crack and peel. You need to put sunscreen on those parts. Some plants will also give a horse with pink skin around his lips a reaction and make him blister. Plus, a white horse will gall easily. (A gall is a sore in the cinch area, usually caused by a rubbing or pinching cinch). A horse with a lot of pink skin out in the sun without any shade will soon get in a bad mood. When the sun burns a light-skinned horse, it hurts the horse just as it would hurt a fair-skinned person.

SKIN CUTS AND GALLS

Anytime you get a skinned place on a horse, like a rub or even a little cut, use something with a large percentage of aloe vera on it. For a gall, use Preparation H. Those are two things you can get without running to the vet. But first, you need to make sure that the cut isn't large enough or deep enough that it needs to be sewn up or cleaned out by a vet. You also need to make sure that if it is a puncture, there isn't anything in it. Anytime the skin is broken, give your horse a tetanus-antitoxin shot. That's in addition to the annual inoculations discussed earlier in this chapter.

VETERINARY CARE

Ninety percent of the horses that hurt don't perform well and would do better if they felt good. Horses save themselves. For that reason, it is our job to keep them feeling as good as possible. Anytime a horse is sore, it doesn't necessarily mean that we won't exercise him, but we probably won't compete on him or work him on the barrels. A lot of times if a horse is just body sore, you can get rid of soreness with light, easy exercise. We do not exercise a horse that is crippled by lameness or other injuries.

Once a year, or when we purchase a horse, we like to have complete blood work done on him, but not just "packed cell volume" or a "blood count." We want to see if he's high in phosphorus, low in calcium or if there's some other abnormality. We think monitoring these things makes a big difference.

Anytime that you have a horse that is nutritionally out of balance, you need to go to an equine specialist, either a nutritionist or a vet. However, many veterinaries are not trained in equine nutrition. Whenever we need advice or assistance, we go to one place for legs, another place for colic, and another place for nutrition. We use a lot of professionals who have specialty areas. It takes some research on your part, and perhaps visiting with some people who are knowledgeable horsemen, to figure out where the best place is to go to achieve the best possible results.

Just as most people don't go to a general practitioner for open heart surgery or knee surgery, we don't go to a general practitioner unless we're taking a horse to get his teeth floated, shots or need advice on deworming. With the high cost of a good horse, we now use leg specialists, colic specialists and equine surgeons, to name a few. We have one vet who does our respiratory and leg work. We have another that we use for colic and another who does our breeding. The best vets are specialists in their fields.

For information regarding equine veterinarians in your area and their specialities, contact the American Association of Equine Practitioners. Or better yet, ask educated horsemen in your area.

CHIROPRACTORS, ACUPUNCTURE AND ACUPRESSURE

It has been proven that chiropractic and acupuncture practices do help some problems in horses. In fact, in 1989, the American Veterinary Medical Association recognized acupuncture as a "valid modality and an integral part of veterinary medicine." Today, many veterinarians are also trained in acupuncture, acupressure and chiropractics. As important as it is, we don't feel that chiropractic help is always a cure, but it makes the horse feel good enough to perform his best, mentally and physically. Acupuncture can actually be a cure for certain things.

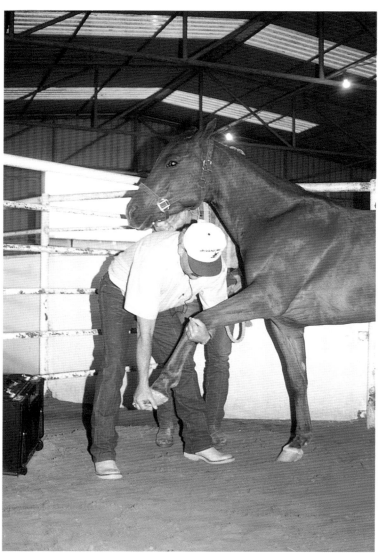

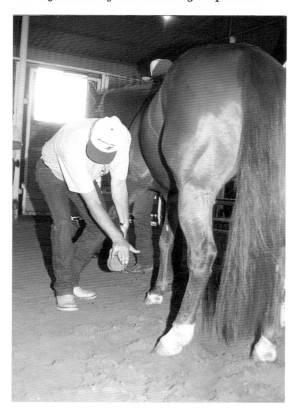

(left) Laird Burke, an equine kinesiologist, adjusts a horse's knee laterally, preparing to adjust the accessory carpal bone located at the back of the knee.

(below) Laird presses down on the heel, completing the adjustment of the accessory carpal bone.

Chiropractic: Chiropractic is the science and art which utilizes the inherent recuperative powers of the body, and deals with the relationship between the nervous system and the spinal column, and the role this relationship plays in the restoration and maintenance of health. Chiropractic work is the use of bone alignment to treat specific or general health problems. Its techniques offer a unique solution for many of the health and performance problems of horses by restoring the function of the back and neck. It is a proven fact that chiropractors have helped barrel horses that have difficulty flexing around a barrel.

As in any profession, there are competent people and not-so-competent people. It's up to you to find out if a particular equine chiropractor knows what he is doing. Obviously, you can't try them all out by letting them work on your horse. The best way to learn which ones are good is by communicating with other horse people. Also, some chiropractors are knowledgeable, but they adjust horses too roughly. An adjustment should be done carefully.

You can palpate or manipulate your horse yourself and see if and where he is sore. Palpate with the tips of your three middle fingers, which are very sensitive. Softly palpate or manipulate the horse's poll and axis, located between the ears. Squeeze it and massage it. You'll get a reaction from the horse if he's sore.

Further information on animal chiropractic and certified animal chiropractors is available from the American Veterinary Chiropractic Association.

Acupuncture: Acupuncture is the insertion of fine needles into specific predetermined points on the surface of the body in order to regulate bodily functions. According to traditional Chinese medicine, there is a flow of energy called "chi," consisting of positive (yang) and negative (yin) components that course through channels in the body called "meridians."

An acupuncture point is an area of the skin one to two millimeters in size located over specific anatomic landmarks on the body that are linked with visceral organs. There are three types of acupuncture points: The primary ones found along the routes of large nerves in the skin and muscle, the secondary points found in smaller nerves and the third type found at small nerve-muscle fibers.

If you palpate down the horse's back, you will find several acupuncture points. If there's a problem there, it's usually a muscle strain or a point that relates to another part of the body. Farther on the back, the problem can be a vertebra that has a muscle pulling on it. The vertebrae of a horse are rather large. But often, horses have spasms in their muscles and relieving those spasms can be done with acupuncture or acupressure.

Information on veterinary acupuncture can be obtained from the International Veterinary Acupuncture Society. This society came into being in 1974, and today its membership includes veterinarians from all parts of the world. This organization conducts postgraduate courses in acupuncture for veterinarians. A list of veterinarians who have completed the course is available from the IVAS.

Acupressure and Massage: Acupressure is the art of utilizing stimulation on pressure points to treat an animal. Acupressure sometimes actually relieves spasms in the muscle without using acupuncture. Veterinarians rarely use acupressure. However, they can use it in conjunction with acupuncture to relieve muscle spasm and pain. Acupressure can be taught to nonprofessionals to augment veterinarian-applied acupuncture.

The best thing is to learn more about acupressure yourself so you can check your own horse. That way, you will know when your horse has a little body soreness. You can acquire a feel for it. Horses tell you they're hurting in different ways. Learn about your horse and what his reactions are. Some horses pop their tail when they're hurting, some move their head, while others just flinch or grind their teeth.

You can also use deep body massage on a horse. Deep massaging of the horse's body relieves muscle spasms, softens muscles, creates circulation, makes the horse relax and feel good mentally, due to the soft feeling in his body. Techniques for deep body massage are not within the scope of this book. You can learn more about this sort of therapy from books, video tapes and magazine articles on alternative therapies. You can also hire a certified massage therapist for horses or take a course in it yourself.

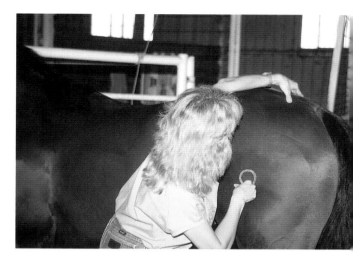

Becky Phillips, a certified massage therapist, massages the hip and stifle muscles to help alleviate soreness. Muscle spasms can also be dissipated with direct hand pressure.

DENTAL CARE

Dental care is important for your horse because it affects his health and his attitude. The two most common and most important problems are the presence of wolf teeth and sharp, uneven teeth.

There are a couple of ways to monitor how good a horse's teeth are. First, if you notice your horse spilling grain out of his mouth while he's eating, there's a good chance his teeth need care. Teeth that are uneven and don't meet one another in an opposing fashion cause a horse to spill grain, rather than chew it properly. Second, you can check the teeth yourself by running your fingers along the teeth, feeling for sharp edges or unevenness. However, be careful you don't get bitten when checking your horse's teeth.

Wolf Teeth: Wolf teeth are little teeth found in front of the molars in the upper jaw. They have no function and the primary problem with them is that they cause irritation with the bit. Most horses dislike a bit banging against the wolf teeth. The best thing to do is have your vet or an equine dentist remove them. The vast majority have shallow roots and are easily removed. However, there are some that are more complicated. If you have an older horse that has wolf teeth and he doesn't react adversely to them, don't worry about removing them.

Floating Teeth: The horse's top jaw teeth sit outside the bottom teeth and they should touch. When the jaw teeth rub and grind on each other, as the horse is chewing, it's just like sharpening a knife. If you live in a really sandy area, where there isn't much grass and the horse is getting a little sand in his mouth, there are times the sand actually works as a whetstone, grinding the teeth and making them sharp on the edges. Also, a nervous horse is prone to grinding his teeth and, therefore, forming sharp edges quicker than a calm horse.

Horses should have their teeth floated on a regular basis, usually once a year. Floating is the process by which the teeth are rasped even. As teeth go through normal wear and tear, they become uneven and develop sharp points. The unevenness doesn't allow horses to chew their food properly, which inhibits proper digestion of the nutrition in the food. Sharp points can also injure the horse's tender mouth tissue and create abscesses. The pain can even cause the horse to stop eating altogether.

Floating also helps the bit work better in the mouth. But what's more important is how it changes the attitude of the horse. If your teeth hurt, you'd be a "gripey" rascal, and a horse would too. We often see attitude adjustments after floating a horse's teeth. The front teeth of the horse need to be just as level as the back teeth. If they're not level, and the bite isn't even, you're in just as much trouble up front as you are behind. Also, canine teeth need to be rounded off and tartar needs to be removed.

FEEDING THE ATHLETE

Just like people, horses need to eat a balanced diet. Feeding only hay and oats is generally inadequate. Most sweet feeds or mixed feeds, especially those containing vitamins and minerals, along with the grains, are better. To determine if your horse's diet is nutritionally sound, have a veterinarian run a complete set of blood tests on him once a year. He can tell you if the horse has any diet inadequacies. To rectify the problem, ask his advice or that of an equine nutritionist.

Horses today are really in trouble because there are so many herbicides and pesticides used on the hay and grain. Also, there are so many additives on the market that it can be quite confusing for most horse owners who only want to do the best for their horses. Read the labels carefully. Don't feed something just because someone told you how wonderful it is or the advertisement sounded great.

Hay: Horses are herbivores, which means their main diet should consist of roughage, such as grasses. We are not proponents of the type of commercially prepared feeds that are called complete feeds. That means you don't have to feed any hay. The feed, usually in the form of a pellet, is said to contain some hay and grain. We believe horses need roughage in its actual form — hay.

We prefer to feed our horses a certain amount of grass hay and a small amount of alfalfa. We need to emphasize again that there is 90 percent consistency and 10 percent variable in anything that we say. Nothing is 100 percent absolute. You need to find out how to feed your horse in your area, because hay and the feed value from it varies greatly from area to area and state to state. Hay grown in some areas is inadequate because of the soil it is grown in.

Nitrogen is needed to grow coastal hay in our part of the country, but you can poison a horse with too much nitrogen. Hay needs to be very lightly fertilized.

We feel the best hay is grass hay without a lot of weeds and definitely with no mold. It also needs to be put up properly, because hay can be put up so dry that it's not palatable and has no nutrition to it. But it can also be put up with too much moisture in it. That will cause mold. We feed coastal hay because we raise it here and have access to it. If you feed alfalfa, use discretion. Alfalfa can be a hot ration with extremely high protein if it comes from certain parts of the country.

Alfalfa varies all over the United States. We know of a top barrel racer that came to Texas with two futurity colts and fed nothing but alfalfa. After she ran out of her alfalfa from home, she bought alfalfa here and her horses started to get thin. The problem was that Texas alfalfa did not have the feed value that the alfalfa from her part of the country had.

Colorado and New Mexico alfalfa is really good, but alfalfa grown in Texas and Oklahoma is dangerous to feed horses because of the high incidence of blister beetles. A blister beetle is a very dangerous insect that is sometimes found in alfalfa hay. Unfortunately, you cannot always see the beetle in the hay. Even if a blister beetle is run through a crimper, it can kill a horse in just a little bit of hay. There isn't anything worse than seeing a horse die from blister beetles. It's like taking a cutting torch to your horse's stomach and burning the lining out of it.

The amount of hay you feed differs by the kind of hay you are feeding. If you are feeding coastal hay, we would suggest feeding three or four flakes per feeding to each horse. If you feed alfalfa, we would suggest the same three or four flakes of grass hay once a day, and one flake of alfalfa once a day, along with two flakes of grass hay.

Grain: Whole grains are not as easily digested by the horse as crimped or rolled grains. Pelleted feeds can expand in the horse's stomach and cause problems from bloat or colic. If you want to feed a pellet, make sure it's the type that breaks down readily upon digestion. Recently we switched to a pelleted feed that does just that. We have found that it costs a little more per 100 pounds, but we can get by with feeding less.

We did feed cracked and steam-rolled oats, which are highly digestible. A mature horse needs only a 10 to 12 percent protein ration, while a growing colt needs a lot higher percentage, from 16 to 18 percent. We don't have two different rations. We use the same ration on both, but we add a supplement to up the protein on our growing colts. If you feed an older horse a high content of protein, he could develop some muscle soreness, so you need to be careful with your supplements and percentage of protein.

We feed a horse that is in training one gallon of feed twice a day. If a horse is just resting and not in training, we would suggest feeding one-half gallon per feeding. When you haul somewhere to compete, feed half of your grain four to five hours before you are scheduled to run. The reason for feeding part of the grain is not necessarily for energy right then, because it doesn't immediately turn to energy, but because a horse is on a schedule, like we are. His stomach has a time clock in it. At our house, we feed at 6 a.m. and 6 p.m. At that time, those horses are going to start fretting and looking for you. They know when they should expect to be fed. That's the reason a competing horse needs a small portion of his grain four to five hours ahead of competition time.

Supplements: Don't go overboard on supplements. With all the different supplements on the market, you can end up with a higher quantity of supplement than you have grain ration. Unless you have a problem that you need to adjust, most supplements are unnecessary. Supplements should be used on horses when their blood count is a little low, or they have an imbalance in their system. If you make sure your horse's rations are totally balanced, he shouldn't need supplements.

If you do use a vitamin, use one that is all natural, not synthetic. You should be able to tell by the label of a product whether or not the vitamins are natural. However, you may have to research the product further to learn the true contents. This can be done by calling the manufacturer or retailer where you purchased the feed.

Also, blood builders aren't necessary for most horses if they have proper nutrition. We do a complete set of blood chemistries on a horse yearly to see if they're low in phosphorus, high in calcium or whatever. Your feed dictates that to a great degree. At the same time, you should have your veterinarian run a thyroid panel to check your horse, because horses can have thyroid problems.

We do use an all-natural supplement, along with a mineral and light salt block that we feed "free-choice" (available at all times for a horse to eat). They can pick up whatever they want of it. It benefits their hoofs and balances the system. It's especially good for pastured horses. We also use a probiotic. These fermentation products enhance the beneficial bacteria in the stomach. Anytime that a horse is being hauled, if the weather changes, if you've used him real hard, if you've given him antibiotics or a dewormer, or done anything to change his daily routine, he might benefit from a probiotic. You give it to your horse orally by squirting it in his mouth or feeding it along with the grain ration. Probiotic products can be obtained from some veterinarians, in some vet supply houses and in some tack shops.

Water: A horse needs to stay hydrated by having an adequate supply of good, clean water. That doesn't mean one bucket in the morning and one bucket in the evening. A horse will drink more than that in the summer, or he can knock his water bucket over, or get his hay or grain in it and make it taste bad. If you have a horse in a stall, hang at least two buckets of water in the stall. We don't like to keep a horse in a stall all the time. All of our horses, sometime during the day or night, are turned out where they have free-choice water from large, clean troughs.

Horses need electrolytes to help them keep hydrated. Electrolytes are usually in the form of a paste or granules that you put in the horse's feed or water. Hot summertime weather especially calls for the use of electrolytes. When you are hauling your horse, he needs electrolytes added to his feed or to his water. It's also important to keep horses from become dehydrated in the winter, because quite often they don't like to drink cold, icy water.

When we travel, we carry water cans, filled with water from home, in the back of the truck. It tastes a little bit like the water can, and obviously it gets hot in the summer, but every time we unload, we can expose the horses to water. Most of the time they don't drink, but if they take just one or two swallows, it is worth the trouble to us. Before competing, we never totally pull a horse off water for four or five hours. We don't give him a full bucket, but we'll let him have part of a bucket of water.

When we're through competing, we don't mind our horses having a swallow or two of water, but they can't have all they want. A horse can founder from too much cold water when he's hot and stressed. We allow a horse to have a few sips of water after a hard workout. Then we walk him three or four minutes, give him three or four more sips, and just keep doing that until he is completely cooled out.

Anytime a horse comes out of the arena and you cool him out physically, that doesn't mean he's totally cool mentally. Until you think he is totally let down and relaxed, don't give him the rest of his rations or all of his water. Allow him to let down slowly. It's important when you come out of the arena to get back on your horse and walk him down to where the action is and let him come down mentally. If he's still wound up and you just jump off him, when he comes back to another competitive situation, he's high mentally. Always use discretion and common sense and your horse will stay physically and mentally sound easier and longer.

"WHEN THINGS AREN'T LINED UP THEY DON'T WORK PROPERLY."

15

Hoof Care and Lameness

There are a lot of barrel horses with bad feet because we have bred so much Thoroughbred into them and bred the foot off of them. In our breeding efforts to improve certain aspects of conformation, we have overlooked or totally forgotten good, sound feet, and so they don't always have good feet. So we need to do what we can to help them.

From the time a horse is born, the balance of his feet affect him greatly. If they are crooked, they can be brought into the correct position through corrective shoeing or trimming within the first eight months of the colt's life. There are small, plastic shoes and heavy, rubber shoes that you can glue on a foal's feet to change how his knees, ankles or hocks sit. We have used this method and redirected problems with a colt's legs. However, this must be done during the first days of the colt's life to get the best results. You can't completely correct the way a horse's joints are set after the horse is six months old.

HOOF TEXTURE AND SHAPE

A good hoof is reasonably hard, yet resilient, with no chips or cracks. There is a glossy or waxy outer coating covering the hoof. The shape is oval with the bottom of the hoof being wider than at the coronary band. This type of shape spreads the concussive forces so a horse absorbs the shock of hitting the ground better. Anytime you want to help the texture of your horse's feet, start from the inside with a balanced diet. There is no substitute or supplement that can take the place of good nutrition.

However, there are some hoof supplements and hoof conditioners that will benefit some horses. Even though there is no one great product to help the texture of the hoof, there are some good ones on the market. Be sure to

(above) To check the levelness of your horse's hooves, support his leg at the middle of the cannon bone and look down the plane of the hoof.
(above right) A healthy hoof should be dry, hard but resilient. and have no visible chips or cracks.

(overleaf) Barrel racing is tough on your horse's legs and feet. Proper care can prevent lameness later on.

put hoof conditioner on the coronary band and on the sole, in addition to the hoof wall. Be careful about putting too much on the soles because you don't want them to be overly soft.

There are several products that your farrier can use for hoof repair. In dry weather, a horse's hoofs get "shelly" around the bottom. With these products, you can go below the shoe nails and it helps reinforce the horse's foot. Above the shoe nails, we use a hoof sealer to help retain moisture in the feet. An overly dry hoof is detrimental to the horse, as is an overly moist or soft hoof.

There's nothing wrong with riding your horse barefoot as long as his feet don't get sore. But it's hard to keep a riding horse barefoot and sound. Anytime your horse gets sore footed, he needs to have shoes. But barefoot is better than improper shoeing. You want the foot level and maintained, whether you are riding a horse barefoot or shod. You can tell if a foot is level by picking it up, grabbing the middle of the cannon bone and letting the foot hang naturally. Look down the plane of the bottom of the foot and you can see if it is level.

THRUSH

It's very important to keep your horse's feet clean because of thrush, which is a degenerative condition of the frog. It is characterized by a black, smelly discharge. It's caused by keeping a horse in wet, unsanitary conditions without cleaning his hoofs and/or the lack of frog pressure from poor shoeing or trimming. To help alleviate this, carry a hoof pick with you and clean out your horse's feet every morning and evening, before and after you ride him and before you load him in a trailer.

SHOEING

There are certain conformation problems that you can counteract with shoeing. For instance, if you have a horse that is back at the knee, you don't want to compound it by leaving a long toe and a low heel. By the same token, you don't want to stand him straight up and chop his toe off and leave a lot of heel. There's a happy medium for everything, which means being balanced and level. For a horse that's over in his knees, don't leave a long toe and a low heel because you will stress the support system of the horse: the deep flexor tendons, the superficial flexor tendons and the suspensory ligaments.

The whole suspensory system at the back of the leg feels like a small piece of elastic rope or cord. The largest and most obvious is the deep flexor tendon. Lying right in front of it, and less obvious, is the superficial flexor tendon. Behind that is the suspensory ligament. The suspensory ligament is probably the one that shows soreness first, is the least of the primary support systems and will heal the quickest.

Damage to the sheath of the tendons, either the deep flexor or the superficial, is a problem. Damage to the tendon itself is alarming. You can help heal the sheath to a certain degree but anytime there is damage, scar tissue forms and does away with some of the elasticity. If you weaken the structure of the horse, he is more susceptible to injury the next time. Consequently, you shoe to minimize the horse's weaknesses.

The hoof is made up of small tubules shaped like straws. Anytime you bend those straws out of shape in any direction, they don't regenerate and grow as well. So it's very important that a horse is shod correctly to be balanced. When your horse is shod, the pastern and the center of the hoof need to be in alignment. Don't completely trust a hoof gauge as they aren't all that accurate. You definitely want a foot level, but the angle of a hoof varies with the skeletal construction of the horse. After you've had a horse shod, and have ridden him for three or four weeks, the toe of the shoe should be worn right in the center. If it's not, the shoe is more than likely off center and it's making the horse stride wrong. When things aren't lined up, they don't work properly.

Every horse needs to be shod according to the slope of his shoulder, the slope of his pastern, the type of foot he has, and his particular problems. Each set of feet are different. It's also important what type of shoes you have on your horse. We like a rim shoe in front and a slick or wide, flat shoe behind on 90 percent of our horses. That doesn't mean a wide, sliding plate; just a regular keg shoe. We want equal rims on both front feet so they fill with dirt, because dirt against dirt holds better than steel against dirt. The front end of a horse isn't as locked against the ground as the back end, and a horse needs to have some stability on slick ground.

We use regular keg shoes on the hind feet. If you put rims, corks or caulks on the back feet of the horse, they will grab, and that's when a horse sores up in his hocks,

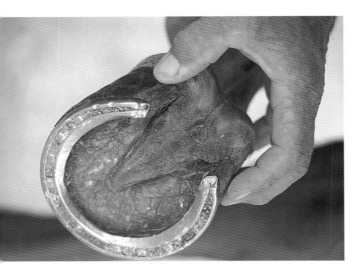

The Wrights use a rim shoe on the front hoofs.

stifle and back. The rear feet of a horse need to be a little bit forgiving. This means that they shouldn't grab and stick as soon as the horse plants his back foot to execute a stop or turn. They shouldn't stick in one place hard and fast. The foot needs to slide between three to eight inches, so that a horse can move his rear end.

The growth of a horse's foot and its condition dictates how often you shoe him, but most horses should be shod every five to six weeks. If we get in trouble, we'd rather have one a little short with his shoes on tight. The biggest problems occur when feet are long, "shelly" and break off, or when the horse loses a shoe. You can pull some hoof wall off or have it break off later. It's very important to keep shoes on a horse that has poor feet.

Overreaching is a common problem with many performance horses and is caused by a number of things. When a horse is hurting and travels wrong, he may overreach (hind feet touching the front feet). The rider sometimes causes overreaching. He could be taking hold of a horse wrong, sticking his leg in the horse wrong or putting his body in the wrong position on the horse's back. Or maybe a saddle doesn't fit. If a horse is narrower behind and wider in the front and he dog tracks (where his front legs are wider apart than his back legs), he will inevitably overreach in a turn.

Don't let your horse go too long between shoeing jobs. If you run your pickup down the road at 65 miles per hour, and it has tires that are going to blow out, it's dangerous. The same is true for a horse. If his feet are too long or unbalanced, and you get into that arena with bad feet, you're in danger of creating soreness or even injury to your horse.

LEG PROBLEMS

Hocks: Typically, the first place that performance horses, such as barrel, reining and cutting horses, have a problem is in their hocks. That joint takes the most stress because of the way horses must use their rear ends when performing in those events. The rear end is the catalyst for turning and stopping. Horses use their hindquarters more than the front end in moving forward, stopping and turning. The first two jumps are what propels the horse forward, then he starts pulling with his front end.

Joint soreness in barrel horses, especially in the hocks, is an area that is all too often overlooked as a source of problems. Pain in the hock joint inevitably leads to soreness in the back, as well as other areas of the body. As the horse tries to compensate for pain, he overuses the rest of his body in an attempt to take the pressure off sore or injured hocks.

Sore hocks can cause a horse to stop rating when he approaches a barrel or stop trying to turn altogether. He may also switch to the off lead behind when loping circles or turning the barrel. When you suspect hock soreness, have your vet take X-rays of the joint to determine the problem. Much of the time, injecting the joint with a medicinal product that targets the area of the joint will help the healing process. There are several methods of treatment and you and your vet must decide which is best for your horse.

Windpuffs, windgalls and leakage: You have to be very selective when you think you are seeing windpuffs as it could actually be leakage of synovial fluid out of a joint. If it's leakage from a joint, or from the stress of a tendon, you need to visit your veterinarian. If there is no heat and no pain, the swelling is probably a windpuff, which isn't much to be worried about.

Windpuffs do not cause lameness, are painless and usually caused from trauma commonly found in the fetlock joint or the flexor tendon sheath above the sesamoid bones. Most hard-working horses have windpuffs. However, anytime you have a little bubble on the knee or hock, you need to have it checked. This is more than likely leakage out of a joint. If a horse that has never had windpuffs before suddenly develops them, check your shoeing. A lot of times it is caused by a horse being out of balance, which places a lot of stress on the foot and leg.

Splints: Splints are calcium deposits on the splint bone of the leg, usually on the front leg. They are caused by injury or a blow to the leg, concussion or stress. Anytime you have the slightest degree of a splint, determine if it's hot or cold. Hot means it's still active with the possibility of enlarging more. A splint is usually hot when it's first forming. After it becomes set, it's called cold. Your vet can help you with any treatment. Splints are a great problem if they are near a joint, tendon or ligament. They can cause friction and, therefore, lameness. They are also unsightly.

When we're exercising our horses, we use splint boots on a horse 90 percent of the time — 100 percent of the time when we are competing. A good splint boot is the cheapest insurance policy you can have. However, anytime you go to the pasture or are going to trail ride all day, splint boots are unnecessary and can even hurt your horse. They can stay on a horse only so long without creating too much heat underneath them. They can also get damp and create friction on the skin and hair.

Always use discretion with splint boots. We don't like any kind of splint boots that are lined with a fleece that absorb sweat and get wet and crusty. We like something smooth and slick, like neoprene. If the boot is made of leather, we keep it oiled and soft. If it's just a smooth, slick material, we keep it clean. The splint boot should be adjusted in the proper position, with adequate tension or tightness to stay in place, but no excessive tightness to create problems.

"DON'T HESITATE TO TRY DIFFERENT EQUIPMENT."

16

Tack and Equipment

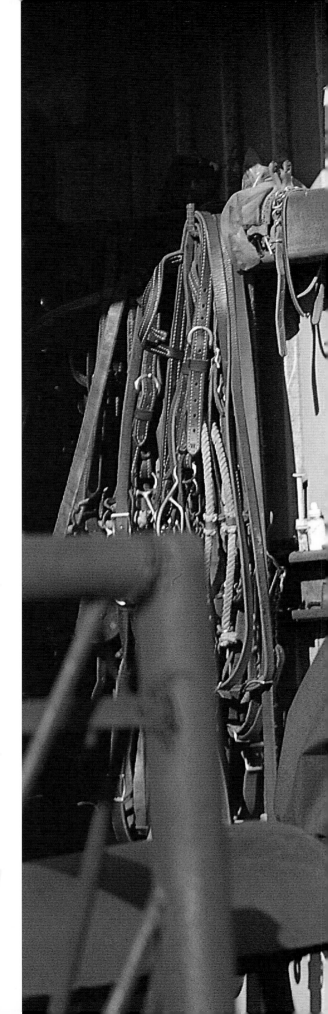

When a professional barrel racer sells a horse for what can be over $100,000, nine times out of 10 the bit goes with the horse. This goes to show how important a piece of equipment can be to your going to the paywindow on your new horse. With competition as keen as it is today, even tack and equipment can make the difference between first and second place.

If equipment doesn't fit your horse, it can make him sore and less likely to "give it all" while he is running. Or if the saddle doesn't fit you, it can put you in an awkward position and you will not be able to ride your best. Equipment that works on one horse may not work on another. Therefore, if you are having a problem that you think may be equipment related, don't hesitate to try different equipment.

SADDLES

The most important feature of a saddle is that it first and foremost fits the horse properly. One way to determine this is by making sure you have from one to two and a half fingers clearance between the horse's withers and the gullet (the part of the saddle that sits over the withers). If the saddle sits lower, it will rub the withers and the constant pressure on the withers can cause soreness and injury. If the saddle sits higher and too far away from the horse, you also have a problem because it leverages against the horse.

While you don't want the gullet to contact the horse's withers, you do want the bars of the saddle to sit down on the horse, contouring to his back, not bridging it. A saddle shouldn't look like it's propped up on a horse's back. The movements of your body as you ride can cause the saddle to sway back and forth, and eventually hurt the horse's

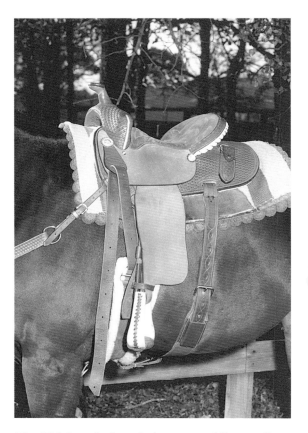

The Wrights design their own saddles to fit a barrel horse and put the rider in the proper position.

(overleaf) Properly fitting tack is a very important part of your training program.

back. Also, if the bars don't lie flat, they can dig into the horse's back, again causing pain and injury. Anytime a saddle doesn't fit a horse, it doesn't matter how it fits you. It's a detriment to the horse, and a big one.

Next, pay attention to how a saddle fits you. You want the saddle to have some ground seat that slopes forward so you can get up in the "jockey" position, yet remain close to the saddle and horse. If the swells sit straight up and down and the horn is high, it traps you sitting down, consequently hindering the horse's ability to perform.

Also, if you're not up front, you're not feeling what's about to happen; you're feeling what's already happened. Remember, however, that "up" doesn't mean standing up. If you're standing up, you're totally away from the action, plus you're prone to being unbalanced in the saddle. Up means you are crouching forward and staying close to the saddle. Sitting "down' is sitting in the middle of the saddle. If you sit back on the cantle, you're out of position.

You want the base of the swell, where it meets the seat, to be cut out so that there is a depression in it where your leg fits underneath the swells. That way, you can get into the forward position. If the saddle is really full under the swells, it won't let you get in the forward position, and you can't ride as balanced in a turn. A channel for your leg cut under the swells gives you a more secure feel in the front end and more balance.

SADDLE PADS

We like saddle pads which are made of natural materials, such as wool. However, wool, which should be next to a horse's back, is getting harder to get. Also, some manufacturers are chemically treating it now, which is not good. So, we also use some synthetic, fleece-lined pads. You want the padding to be adequate, but you don't want so much padding that it puts you up and away from the horse; it makes the saddle slide around on the horse's back.

If you have hair that mats up on your saddle pad, use a big rubber curry comb and brush the underside in a circular motion in order to remove any matted hair. You can wash pads, but don't wash them in a detergent because many detergents are abrasive to a horse's skin. Just wash them with clear water. You can also use a high pressure hose or even a garden hose with a pressure nozzle to wash off dirt, hair and grime.

CINCHES

A horse has to breathe comfortably to perform properly. It's important that you don't cinch him too tightly with a cinch that has no give. The give is in the latigo and off-side billet strap between your saddle dee rings and your girth dee rings. You don't want your girth pulled up against the dee ring because you want the horse to be able to breath. Therefore, it's important that the girth be the proper length for that particular horse. To determine if the girth is the proper length, make sure that the cinch ring sits even or slightly above the horse's elbow on each side. This is a kind of trial-and-error situation on every horse. Pull the cinch up on a horse just enough so that the saddle doesn't move sideways or back, but don't cut him in two. Pull the cinch up enough to be safe, but not so much that it hinders the horse.

The material that your cinch is made of is very important. We don't like mohair-strand cinches because each strand isn't of the same tightness. One will pull a little differently from another one. Also, if a mohair cinch gets wet, it has a different tension than when it's dry. And in cold weather, it also has a different tension than when it's hot. Consequently, it also changes while you're riding. We like something with some width and good shock absorbency to it.

The cinch that we prefer is a nylon cinch with wool fleece lining. On a nylon cinch, the pressure stays consistent and the cinch doesn't stretch. However, one of the bad things about a fleece-lined, nylon cinch is that you have to be careful not to let it get covered with dirt and burrs. It's hard to ride with it on in the pasture without collecting burrs. You need to keep your cinch clean and with a lot of life in it. If it ever gets compressed and worn out, get rid of it.

If you have more than one horse and want to ride them properly, fit every horse with his own cinch and a pad. When your horse sweats, it will soak into the pad and when it dries, the pad will contour to the back, making the saddle fit that much better. Where the cinch ring sits is important and the correct length of a girth helps determine this too. If it sits too high, it moves the saddle back where you don't feel what's about to happen. Plus, if it's back so far on a horse that he's dragging you, you might as well be sitting back there in a buggy, driving him around the barrels.

BREAST COLLARS

If you have a horse with a really good back and withers, you don't need to use a breast collar because a breast collar does restrict some action in the shoulders. But we probably haven't had more than two horses in our lives with a perfect back. Therefore, it's much better to use a breast collar to help keep your saddle in the proper place on the horse's back.

The breast collar needs to be a very smooth, high quality leather. It should be two to two and a quarter inches in width and very soft. Don't use one that is only a half an inch thick with a bunch of little rawhide knots braided on it. If it's thick enough and creates enough specific pressure spots on the horse's shoulder or chest, it could cause him to slow up or quit running in an attempt to keep from making contact with it. Also, the breast collar needs to have a leather hobble in the front that attaches to the cinch between the front legs. This holds the breast collar down so it doesn't catch your horse in the base of the neck. If it does, it will restrict him in the middle of a turn and possibly even interfere with his breathing.

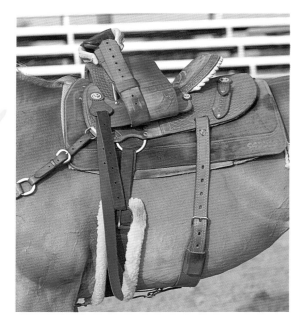

Ed and Martha like a nylon cinch lined with 100% wool fleece.

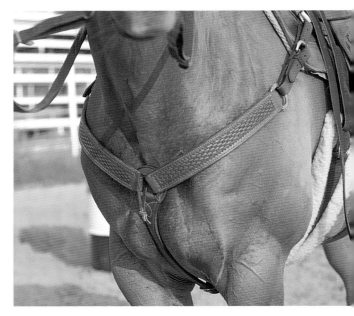

A breast collar should fit properly and never hinder the movement of the shoulders.

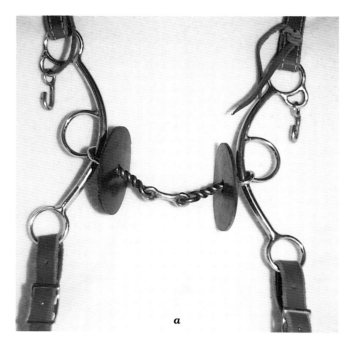

a

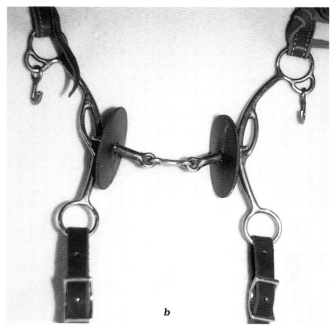

b

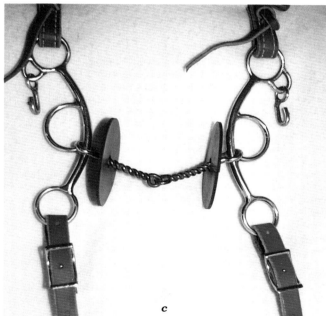

c

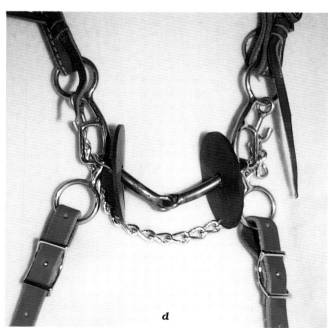

d

a) three-piece twisted long shank.
b) three-piece smooth long shank.
c) two-piece twisted medium shank.
d) two-piece smooth short shank.

BITS

If we had to, we could use the same bridle on every horse that we ride. That is if we and our horses had enough education. Bridles and bits are nothing but shortcuts or aids in control. Anytime you are having a problem, slow down and educate yourself and your horse a little more. That's better than getting a bigger bit. Occasionally, however, there are horses that, through the lack of training or improper training, have a bad attitude about life, and sometimes you can change that attitude by changing the bit. Also, sometimes we use different bits to "freshen" a horse's mouth and mind, especially if we've got several competitions or futurities in a row.

For each horse, we try to keep at least three bits that work for that horse. That way, if we go somewhere and our horses are starting to get mentally tired and dull, we can usually switch bits and make their minds sharper. Or if they are getting a little "pushy" into the bit, we can switch bits and get a little more respect, just because it's a change. One important thing to remember is that the smaller around the mouthpiece is, the more severe the bit. However, always remember that a bit is no worse than the hand that is pulling on it. The four mouthpieces we use are: 1) a three-piece twisted snaffle, 2) a three-piece smooth snaffle, 3) a two-piece twisted snaffle and 4) a two-piece smooth snaffle. All of our bits have bit guards (a circle of rubber on each side of the bit) on them to alleviate the pinching of the horse's mouth.

Three-piece twisted snaffle: This bit breaks twice in the mouthpiece, which puts pressure on the tongue and bars of the horse's mouth. Unless you have good hands and your horse is well-educated, you will exert excess tongue and bar pressure and give your horse a trapped feeling. Most people are better off working on the corners of the mouth, by seesawing the reins and moving the bit from side to side in the mouth. But remember, the twist in the bit makes it bite the corners of the mouth a little more when you seesaw it back and forth.

Three-piece smooth snaffle: This three-piece bit is an easy bit. It doesn't "bite when it talks" because it has a larger, smooth bar, unlike the twisted snaffle. This bit has a lot of feel to it. Also, you have separate side controls. This means that with your right hand, you're turning right. With your left hand, you can put the reins against the horse's neck, keeping him from "floating" (moving) his shoulder or his rear end out, and also keeping his hindquarters under him.

Two-piece twisted snaffle: This is the most severe of the mouthpieces. It's a small, twisted wire which breaks only once in the center. The single break in the middle of the mouthpiece makes the center of the bit rise into the roof of the mouth when the reins are pulled.

Two-piece smooth snaffle: This bit is more severe than the three-piece smooth snaffle, because it has only one break in the center, but it's certainly less harsh than the two-piece twisted snaffle.

With each of these mouthpieces, depending on the horse's needs, we have the option of short, medium or long shanks, which have a lifting or leverage effect on the horse. By lifting, we mean we can control the horse's shoulder, rib cage, neck and head position with greater ease. When we ask for a desired position, the communication is more clear to the horse if a lifter-type of shank bit is used. The lifter or leverage effect of the shank depends on the distance between the mouthpiece and the headstall. The more distance, the more lift you get on the shoulders, as well as control of the rear end. We also like a slight gag effect, which means that the mouthpiece moves up and down on the shank.

(top) short shank. (center) long shank. (bottom) ring snaffle with no shank.

A long shank on a bit does not make a bridle severe. This is used on many of our horses with a three-piece, twisted mouthpiece. Most horses and riders get along with this bit because the rider has a lot of feel, and medium, not severe, control. A bit with a shank works not only on the horse's mouth, but also on the poll, and the poll is a sensitive area.

There's about an 18-inch area in front of your saddle horn where you naturally want your hand most of the time. But, there are times you need to lift your hand. When you elevate your hand, the bit comes up off the tongue and bars and gives the horse some relief. Your bit then works high on the cheeks and even comes up to the top of the mouth to a certain degree.

When you lift your hand, you can move a horse forward and place his shoulders where you want them a little better, while you are going forward. If you're in the middle of a run and you need a little more forward motion, pick up your hand to control the horse's shoulder. That way, you keep the horse's rear end under him, but still send him forward a little more by picking your hand up. But you need to feel his mouth and "give and take" with the reins.

Combination Bit: On a combination bit, the mouthpiece and the nosepiece should work simultaneously. A lot of the combination bits on the market have some mouthpiece effect and then after the mouthpiece takes effect, there is a hackamore effect. A lot of people think a combination works so well because it's working on the nose and the mouth simultaneously. Most of the time, however, that's not the case, as very few of them apply pressure to the two areas at the same time.

A combination bit works on many horses because it has a cavesson effect. A cavesson is what you use to tie a horse's mouth shut, and that's what the nosepiece and the curb are doing, tying the horse's mouth shut. When the horse can't gape open his mouth, he has more feel on his face. We never like a horse's mouth tied solidly shut with a cavesson or a combination bit. If you have a cavesson on a horse, you want it adjusted to where he can open his mouth a little and get some relief. It gives him a really claustrophobic feeling to have his mouth tightly tied shut.

Mechanical Hackamores: With a mechanical hackamore, no matter how well it is constructed for the purpose of turning a horse or making him supple, the hand has to give and take on the reins for it to be effective. That's why we sometimes call it the "stop-and-go" bridle. If you pull on a mechanical hackamore, it makes a horse stiffer. However, if you give and take with your hands, you can make him supple.

The mechanical hackamore has to hang low on the bridge of the nose, just above the soft part of the nostril to be effective. If it's below the bridge of the nose, it usually causes a horse to elevate his head and front end and push into your hands. But if you keep it low on the bridge, it will allow your horse to be supple, flex at the poll and the withers, and be soft laterally.

The nosepiece and mouthpiece work simultaneously on a combination bit.

The noseband of a hackamore should hang low on the bridge of the horse's nose.

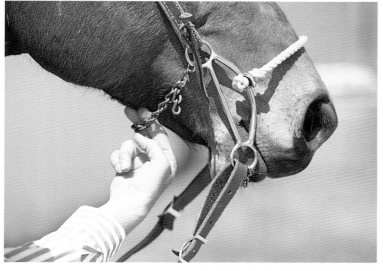

If adjusted properly, you should be able to slide two fingers under the curb chain on a hackamore.

It's important how you set your curb strap on a mechanical hackamore. If it's extremely tight, you can't pull on the reins much before you get a reaction from the horse. The curb strap contacts the horse's chin groove too soon after you pick up on the reins. There's no time to signal the horse slowly. He'll feel the pinch immediately. You want your shank to swing back toward you one-third of the way, so you can lead your horse's head around the turn and position his body. On the other hand, you don't want the curb too loose either. If it is, and the shank comes back toward you half way or more, that's too far. Then you're just getting a direct pull, which is no better than a single noseband sidepull.

The most severe mechanical hackamore we have has a long shank and a swivel in the cheek, where you can turn it out, and turn a horse around a barrel with it very effectively. How you set the curb on this, though, is very important. If the curb is loose, this hackamore will turn out, just like a chain gag when the headstall is too loose. You have to set the curb properly.

We really like to use the straight-shank mechanical hackamore, or the one we consider to be of medium effectiveness. It has a lifter effect because the noseband is lower on the shank than the curb chain which it makes a horse tuck his nose and pick up his shoulders a little better. We like a mechanical hackamore on a horse if we're going to take him to the barrels and bump him with our hands. When we scold a horse, we don't yank on him. We just bump. If we bump, we stay out of his mouth. If the horse isn't listening and just goofing off, and he needs a little scolding, we'll put a mechanical hackamore on him.

Free-wheeling gag bit: The free-wheeling gag bit is usually a D-ring gag bit on a rope. The rope goes over the horse's poll, through the gag bit and up to the rider's hands, with the same rope being used as reins. There is no stop for the gag; it just goes up the side of the rope. We don't usually use a free-wheeling gag bit, because it makes a horse want to lift up and push into the bit. It just doesn't give you a good feel of the horse's mouth.

However, anytime you place any type of a gag in the horse's mouth, the curb chain needs to be loose. You need to pick the bit up until the horse looks like he is grinning. If he's not grinning, it will let the shank turn out, which makes a horse pushy in the bridle. If it's a half inch down and not touching him, and you suddenly pick it up, you will have no contact. When you pull a little more, the contact will come suddenly and will startle him. On the other hand, if he's laughing (the corners of his mouth are pulled up too tight), he's hurting, and the gag bit is too tight. To check to see if the curb chain is just right, put from one to three fingers between the chin and the chain. With three fingers, you're using no curb at all. With two fingers, it has a slight effect, and with one finger, it's snug.

If we are riding with a snaffle bit on a horse, we often ride without a curb. We do this because sometimes a horse can get a claustrophobic feeling if the curb is not adjusted properly. With the curb off, you are just working on his cheeks. If you hurt the tongue and bars, you kill the sensitivity in the horse's mouth and it will never come back. If you get the cheeks slightly pink, they'll heal and come back 100 percent.

Solid Mouthpiece Bits: We use some bits with solid mouthpieces, but you can't keep a horse as supple and soft in the shoulders with those bits. Some horsemen can ride a solid-bar, grazing bit with locked cheeks and still keep a horse soft. But they're not trying to beat the clock, going at high rates of speed, turning and running again. They can go across the arena fast and do turning maneuvers fast, but they're not doing it to beat the clock. Also, they're usually not allowed to use two hands. If they do use two hands, they will usually ride a bit with a broken mouthpiece and no shanks.

We like free-swinging cheeks and broken mouthpieces. They give you separate side control with your hands. Your seat, legs, calves and heels all enter into this, but your hands need to be doing two separate things if needed.

Eventually, we want to move to a one-handed ride and then eventually put the horse "on his honor." When we're training or competing on a green horse, he needs control and assistance from us, not forceful or constant control, but help from us when needed.

Correction bit: What we call our correction bit has a solid bar in it and bites quite a lot, but it does have a port that allows a little tongue relief. The way the port sits in the mouth of a horse, when you pick up on it, it rises into the roof or soft palate of the mouth, a rather sensitive area. The shank swivels out for lead-around-the-corner effect. However, with the solid mouthpiece, you get a stiffer kind of horse. This bit doesn't look like it has a very long shank, but due to the way it is shaped, it's harsh and it bites. Plus, the distance between the headstall and the mouthpiece is not very great, therefore, you get a vise effect. This bridle doesn't do much good if the curb isn't fairly snug.

Easy stops: Easy stop bits are a little different kind of bridle. We have them, but we don't use them on too many horses. The easy stop works under the horse's chin, on the jaw bone. If you bump a horse under there six or eight times, kind of hard, he becomes very sensitive and sometimes overreacts. But that's the goal, to get those older, dead-mouthed horses to respect you a little. However, the easy stop doesn't encourage much flexibility in a horse, and you have to decide how much you're willing to sacrifice in the way of suppleness, in order to get some stop and slow down. This bit also causes a horse to execute much flatter, sometimes even stiff, turns.

REINS

Reins should be made of soft, flexible latigo leather, which has some give to it and allows you to feel the horse's mouth. You should never pull on the mouth, just be able to feel it through your wrist and little fingers. Reaction time in your wrists and little finger is fast, which, in turn, allows you to communicate with your horse quickly. If you were to pull the reins with your whole arm, your reaction time would be slow and your feel, or pull, would be too strong. You just want to feel the horse's mouth, not pull on it. If you pull on a horse long enough, eventually his mouth will become hard and insensitive. A rider constantly pulling on the bridle reins encourages a horse to lean into the bridle, and that is a form of resistance. You want a horse that is light and responsive to your hands.

When you're riding the barrel pattern, we suggest using a single, roping rein. Your hand should be from one to three inches in front of the saddle horn. Leave a little longer rein if you're riding in the pasture. You should have contact if you pick up half an inch, and you should be able to turn a horse loose if you release half an inch. Also, you need separate control of each rein when your hands are working independently of each other. If your reins are too narrow, your communication line to the mouth is too thin, and you have to hold too hard to control a horse. If they are too wide, you can't feel the mouth.

A correction bit has a solid bar but also has a port which allows a little tongue relief.

We don't like to use nylon equipment. Nylon reins don't give you any feel of the mouth, no matter if they are braided or sewn together flat. And nylon headstalls move around and don't allow consistent and proper positioning of the bit in the mouth. Headstalls should also be made of soft latigo leather. When you put a headstall on, you want it to stay in place.

TIEDOWNS

Most of our horses aren't ridden with tiedowns. We feel that a tiedown is usually more for the person than the horse. It's a short cut and a means of forcing a horse into a position, rather than educating a horse to get into that position. If your hands react quickly and hard, or you kick the horse too hard, his head is going to come up. So rather than putting a tiedown on a horse, we suggest educating your hands.

However, there are circumstances under which we do suggest a tiedown. It's good for a horse which has a tendency to raise his nose or his head, and move away from your hands. Without it, he could get out of doing what you are asking him to do. Also, there are some horses which use tiedowns for balance, but they don't have to be adjusted very tightly. Anytime you put a tiedown on a horse, you take away from his athletic ability because he's unable to be as natural moving as he would be without one. If a horse is able to get his head up in the correct position, he can get his rear end down easier. If you tie the head down, then you force him down on his front end, and you can't get his hind end underneath him as easily.

We like to use a tiedown that is just a soft rope or a leather noseband. The strap that goes between the front legs needs to have some width to it and kept soft. We set a tiedown so that it will just balance the horse. He won't be able to raise his head so high that he loses sight of the ground and where he is going.

MARTINGALES

When barrel racing, there is only one martingale that is advantageous to you — the German martingale. Like a tiedown, it's a short cut, but it's a good one. There are draw reins that you run through the cheek of the bit and down to the cinch dees, or a standing martingale or yoke that attaches to the breast collar or cinch ring and has a couple of rings to run the reins through. But these are always pulling on a horse; there's never a stop to it. When the horse elevates his head higher than you want, the martingale goes into effect. When he lowers his head where you want it, the martingale is totally loose and your direct rein goes into effect.

When using the German martingale, you can run barrels much easier than you can with other martingales. But anytime that you use a martingale, you still have to use discretion. It gets a divided pressure from the hands. Be careful not to set it too low. If you set it where it engages the horse lower than he is naturally built to travel at high

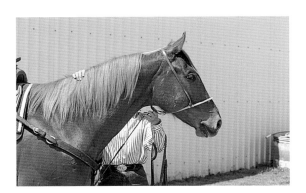

Adjust the length of the tie-down so that you can push the strap up to the horse's throatlatch.

The German martingale should be adjusted so that the martingale only engages if the horse raises his head out of the proper position.

rates of speed, it's always drawing down like the draw reins or standing martingale. That's not your goal. Your goal is that, when your horse elevates his head too high, the martingale will help draw it down. When he lowers his head to the proper and correct position, you have a direct rein on him.

OTHER BARREL RACING AIDS

Boots: Barrel horses need protection in the form of leg boots — splint boots, bell boots and back boots. You may have a horse that you have run a thousand times without ever hitting himself. But the one time he does, he might cut a tendon in two. And usually a horse is going to hit sometimes in the middle of the turn. Nothing you can do in the way of shoeing prevents overreach problems in the middle of a turn.

Having a horse shod level is very important. When you're running in loose or unlevel ground and your horse digs into the dirt, he may hit himself unless you use protective boots. Back boots are like a skid boot. If your horse burns his fetlocks while he is turning and setting, he's going to quit setting and turning unless he's got lots of heart and try. Boots really help avoid those problems.

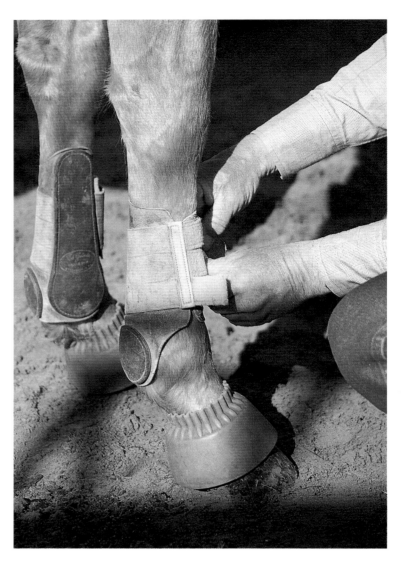

Splint and over-reach boots should be used during slow work as well as competition.

Spurs should be very dull and always used with discretion.

If you're located where it's really hot, like Texas in the summertime, the boots need to come off of your horse's legs every 30 minutes to let the leg dry and to clean the boot out. If you leave boots on too long, you will get your horse's legs raw and sensitive, as dirt works just like sandpaper under a boot. Therefore, right after you work your horse, you should get the boots off and brush the horse's legs clean.

Spurs: There are only two ways to ask a horse to get into position or to ask him to run if he's not sensitive in the sides. Those two ways are with spurs or a quirt. Most people use spurs like weapons. If that's the case, those people should be licensed to use spurs, but not with a 10 day waiting period. There should be about a 15 year waiting period.

Spurs need to be very dull and blunt. They need to be used on a horse just like you would want someone to tap you on the shoulder. You ask a horse to get into position with spurs; they're not to be used to make a horse run. You make a horse run through education, or maybe by tapping him lightly on the rear with a bat or quirt. Anyone wearing spurs needs to be very ambidextrous. A lot of people do a lot with one foot and don't think they're doing it with the other one. In reality, they are using both feet simultaneously. That is definitely a negative. If you mean to spur only one side, make sure you are not using both spurs instead.

Spurs should be used with a lot of discretion. We use spurs behind the flank cinch quite a bit. Anytime you reach back in the flank with your spur, a horse will put his rear end under him. But the first time you stick your foot in a horse's flank, he might buck you off. So until he's educated to your feet, don't put your feet back in his flank.

Bats and quirts: We don't recommend holding your bat in your mouth or hand while running barrels. The reason is that a bat can make your horse move to the wrong position. If you have a bat in your hand or in your mouth, it sticks out just like an aerial. A horse looks at it and he knows that eventually he's going to get smacked with it, otherwise you wouldn't be carrying it. But if you do use a bat, we recommend hitting the horse with a balanced hit. If you hit a horse on one side, he's going to move away from the hit or into the hit. Of course, here anything can happen, but if you had 100 horses, 90 percent of them would move over away from the hit.

Never hit a horse on the shoulder. There are some people that hit a horse on the shoulder with the palms of their hands and that's fine. But when you hit the shoulder with a quirt, your timing needs to be perfect and most novice riders don't have perfect timing. Also, in most associations, including the American Quarter Horse Association (AQHA) and American Paint Horse Association (APHA), hitting a horse in front of the cinch is grounds for disqualification.

Over-And-Under: When competing, we usually use an over-and-under which is a rope with one end locked in place on the saddle horn. When you need it, you don't look down to get it, you know where it is so you can just reach for it. The over-and-under should be long enough to fall 12 to 18 inches under the skirt of the saddle. It shouldn't be so long that you can hit a horse on the soft flesh of his flank or belly. The over-and-under always needs to have wide, leather ends so that it just stings and doesn't hurt the horse. The best way to hit with an over-and-under, or a bat, is with a balanced, even hit. A short quirt looped over the two fingers, between the index finger and small finger, and out the back of the hand, is a super incentive. The length should reach six to eight inches under the skirt of the saddle. A soft leather end should be used on the quirt.

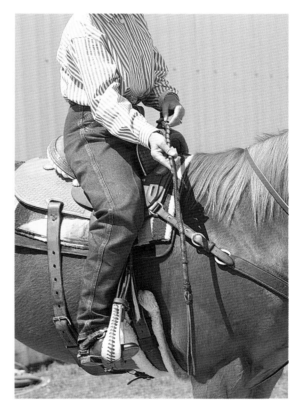

The proper length of an over-and-under is important. Too short is ineffective and too long can actually take run away from a horse.

"COWHAND BREEZE WAS THE GREATEST HORSE I'VE EVER RIDDEN."

17

Great Barrel Horses We Have Known

COWHAND BREEZE (1966-1972)

Cowhand Breeze, nicknamed Breeze, was the greatest horse I've ever ridden. I also feel that he was one of the great horses of all times. He was a black gelding that my dad bought for me as a coming two-year-old at a Quarter Horse Sale during the Houston Rodeo. He was raised by Don "Cowboy" Mehrens, who owned a lot of King-bred horses. He was sired by Chief Breeze by King Breeze, and was out of Cowhand Lady by Cowhand Joe.

We sent him to John Carter, a cutting horse trainer, to break. After he rode him about 120 days, we brought him home and I was literally running barrels on him within a month. At that time, he was still only two years old. He was definitely meant to be a barrel horse, because at that point and time, I was very uneducated as far as riding and training a barrel horse. He did a lot of it on his own.

In his four-year-old year, I won the two-run Ladies Barrel Race at the Texas Barrel Racing Association Futurity on Breeze. At that time, I didn't know what a futurity was, and he had already won over $15,000 and the TBRA Futurity was held for horses that had never competed in barrel racing competition prior to the Futurity.

That same year, Loretta Manual took Breeze to the National Finals Rodeo and won five go rounds on him. The next year, 1971, I again won the Ladies Open at the TBRA Futurity. Also, I was on the Tarleton College Rodeo Team and was named the National Intercollegiate Rodeo Association National Barrel Racing Champion at the Finals in Bozeman, Montana. The team also won the Southwest Regional title.

I was only 18, but I got my WPRA card and qualified for the National Finals Rodeo. On my first run, he turned before we got to the second barrel. But after that, he won six of

(previous pages) Cowhand Breeze, a 1966 black gelding by Chief Breeze by King Breeze out of Cowhand Lady by Cowhand Joe, was the greatest horse Martha's ever ridden. She also believes that he was one of the great horses of all time, even though he never reached his prime. He died of colic in the spring of 1972. (Kenneth Springer Photo)

the 10 go-rounds and placed in nine. I not only won the Rookie Of The Year, but set a new record for money won. In this day and age, it would be peanuts, but at that time, it was a record.

Breeze had marvelous turns, was really fast and was so catty. It's so difficult to describe his turn. He was literally running in the front end and setting in the back end in the turn. He cowed a barrel like a cutting horse would work a cow. He had such forward momentum in that turn; he was so low to the ground and so collected. He could stand up on glass. He just had no ground problems at all.

In fact, if you took him to a rodeo, all the other horses would run on dry ground, but when it came his time to run, there would be 10 inches of mud. He could still win first and outdistance the competition. Ground didn't make any difference to him.

Breeze never reached his prime as he died of colic in the spring of 1972, at the age of 6. I rode him the first half of the year and then borrowed a horse, Liz, which I rode the second half. After Breeze died, I had a full sister to him which I planned to take to the 1972 futurities, but she too died, two weeks before the futurity.

BILLY BARS BUG

Billy Bars Bug was a futurity colt during his four-year-old year in 1979, and was probably one of the most successful futurity colts we've ever had, up until the last few years. There's a funny story behind him. We bought him off the track and he had a good pedigree, being sired by Jerrys Bug and out of a daughter of The Ole Man. When we went to make a barrel horse out of him, he just acted like he wouldn't pan out to be anything at all.

In fact, we tried two or three times to sell him cheap. We offered to let anyone pay him down, pay him out — anything. We had him entered at the Old Fort Days Futurity in Fort Smith, Ark., strictly because at that time they had a sale for horses that were entered in the Futurity. So, we entered him in the Futurity to put him in the sale.

Then, about a month before the futurity, it was like the light came on for this horse. He became so consistent. He was not a race horse by any means, as far as speed was concerned. If a faster horse made a good run, you could outrun him. But his good point was that he was so consistent; he made the same run every time.

We took two horses to Fort Smith that year. Billy Bars Bug, nicknamed "Pokey" and Robyn Leatherman's horse, Fairbert. Fairbert ran the fastest qualifying time and finished third in the Finals. Pokey ran the fifth fastest time of the Futurity trials, but I tipped a barrel over. Everybody wanted to know how much we wanted for him. At one point, we had three people standing there trying to buy him. But we had decided by then, that we just didn't want to sell him.

That year, we took Billy Bars Bug, Top Bar Bug and Pas Reb Pas to the Colorado Barrel Futurity. "Pokey" won both go-rounds and the average, while Top Bar Bug finished 12th in the average. I went on to win both go-rounds and

Billy Bars Bug, a 1975 gelding sired by Jerrys Bug and out of a daughter of The Ole Man, ended up being one of Martha's most successful futurity colts. They tried to sell him cheap during his early training because he didn't act like he would amount to anything. (Kenneth Springer Photo)

the average at the Indiana Futurity. Also, in 1979, Billy Bars Bug was named was Reserve Champion in Junior Barrels at the AQHA World Show.

During his Derby year, 1980, I won the Old Fort Days Maturity riding Billy Bars Bug, and won the Old Fort Days Consolation on Orange Bug. I won the West Texas Derby on Billy Bars Bug and finished ninth in the average riding Orange Bug. I also won the Derby at the Colorado Spring Futurity and finished second at the Rodeo Of The Ozarks Maturity. I finished sixth in the Colorado Futurity, riding Orange Bug.

The year 1980, was a good year, as during a 10-day period, from May 23-June 1, I made seven runs on "Pokey," never placing lower than third on any of them, winning $7,122. After his futurity and derby years, I rodeoed on him about four years, winning from $40,000 to $45,000 on him.

JETTA JAY

I won 1980 Texas Barrel Racing Association Futurity on this mare, while she was owned by Janice Sanders. At that time, it was the largest, most prestigious barrel futurity. Janice brought her to us straight from the race track in the middle of July in 1980. The TBRA Futurity was the end of September.

She had tremendous ability, but had a mind of her own. She shook her head at you and would pop her tail, letting you know she was not happy with what you were asking her to do, but she could outrun anyone that wanted to hook up with her. Ed roped on her in the mornings, and I ran barrels on her in the evenings. That's how we got her to go in such a short period of time. Sadly, she was killed in a horse walker accident at a young age.

RAMBLING RALLY

Rambling Rally, sired by Rally Racer, was also out of a daughter of The Ole Man. During his futurity year, in 1983, I was Reserve Champion at Fort Smith and then went back the next year and won the Derby in 1984. I also picked up paychecks from the Chisholm Trail, West Texas, Budweiser, OCA, Rodeo Of The Ozarks, North Texas and Champion of Champions Futurity. In the period of time that I rodeoed on him, I won from $110,000 to $115,000 on him.

Shy-Anne Bowden, a youth, is riding and winning a lot of big youth and open barrel races on him today. He's not a building horse, but he is fantastic outdoors. The farther it was to the first barrel, the better we liked it.

MAJOR MOVEMENT

Major Movement, sired by Bunny Bid and out of an own daughter of Jet Deck, was a 1986 Barrel Futurity Horse. We bought him from one of our customers, Judy Carey, who had ridden him in the futurities, but hadn't gotten along too well with him. He was really sway backed. Probably the reason for his lack of success was the fact that no saddle would fit him. After making a mold of his

Rambling Rally, a 1979 gelding sired by Rally Racer, out of a daughter of The Ole Man, was ridden by Martha to the Reserve Championship of the high-paying 1983 Old Fort Days Futurity in Fort Smith, Ark. The pair returned in 1984 to win the Derby. (Kenneth Springer Photo)

Major Movement, a 1982 gelding sired by Bunny Bid out of a daughter of Jet Deck, was really sway backed, and the Wrights had a problem getting a saddle to fit him. After making a mold of his back and have a saddle especially made for him, Martha had tremendous success on him, winning close to $100,000. (Kenneth Springer Photo)

back, we had a saddle especially made for him. That made a tremendous difference in his attitude, disposition, and desire to run barrels.

In 1989, I finished second in the Sweepstakes at the World Championship Barrel Futurity, taking home close to $11,000. In 1990, I won Sweepstakes at the Lazy E Futurity, finished third at the big Sweepstakes held during The Texas Barrel Race in Fort Worth, and was Reserve Champion of the Heart of Oklahoma Sweepstakes.

I won the WPRA Texas Circuit on him and that same year, I qualified for the NFR on him, without leaving the state of Texas. However, I chose to not go to the NFR, because I couldn't keep him sound. Also, I had a good futurity horse, Danish, and the Oklahoma City Futurity was on at the same time. I felt that it was a better business decision to go to the Futurity with Danish, rather than try to make 10 runs in Las Vegas.

In 1992, I won the American Novice Horse Association Finals, taking home a two-horse trailer, as well as finishing fourth in the Texas Circuit Finals of the WPRA. Altogether, I won between $85,000 to $100,000 on him. In fact, we never took him to Houston when I didn't win something on him.

DANISH

Danish, a 1989 futurity horse, was sired by Fols Native (TB), out of a Drill Bar mare by Lee Bar by Three Bars. When we bought him off of a track in Kentucky, he had a 99 speed index. The people we bought him from had brought him to us to try on their way to a Paint race futurity with a Paint filly. When they let him out of the trailer, he had three inches of lather on him and he was a basket case. This Paint filly that was supposed to run in a race was just as calm and cool as she could be.

When they unloaded him at Trinity Meadows, Ed saw him and said that he didn't think he wanted him. But they didn't have a stall for him, so we told them we'd keep him until the Paint filly went home. Ed loaded him in the back of our six-horse gooseneck trailer. By the time he got back to

The Wrights didn't think they wanted Danish, a 1985 gelding sired by Fols Native (TB) out of a granddaughter of Three Bars, when they first saw him. But he went on to win over $30,000 for them in futurity and derby competition. He became a great open horse and today is owned by World Champion Barrel Racer, Sherry Cervi. (Kenneth Springer Photo)

Stephenville, he had pawed all the mats back underneath him. He was so aggravated. He was three at that time.

When we turned him in the pen, he had his head stuck nine miles in the air. But we got to watching him, and he could run up in the corner, stick his rear end underneath him, and turn around quicker than anything we'd ever seen. So, we decided maybe we'd better ride him and make sure we didn't want him before we let him get away.

After we rode him, we really, liked him. I don't know what happened in the trailer ride from Kentucky to Texas, but it was not typical of him at all. He was pretty low key, easy to haul, easy to get along with, and had lots of personality and disposition. He was kind of a clown. He was one of the most fun horses I ever had to ride. He was so aware, so bright and ready to go and do something. He was a willing horse and just one of those that you knew you were going to win something on when you got on him and you knew it was going to be fun doing it.

During his futurity year, 1989, I won the North Texas Futurity on him, and placed at several other futurities, including the Lazy E, San Antonio, Texas Arena News, Janet Meyers, Pineywoods and World Championship Barrel Futurity. I tipped over a $52,000 barrel at Fort Smith on him in the Finals. I ran a 17.1 and it was won with a 17.3. I caught the first barrel leaving it. But after that, I went on to win over $32,000 on him in futurity and derby competition. He became a great Open horse and World Champion barrel racer, Sherry Potter Cervi, now owns him.

DEVILISH HONOR

Devilish Honor, nicknamed Elvira, a 1992 futurity colt, she won the Texas Finest and Ardmore futurities. She also made the finals at Fort Smith and The Texas Barrel Race. During her Derby year, she finished seventh in the average at the World Championship Derby. Altogether, she won over $15,500 in aged-event competition.

JETTA C LEO

Sired by Jet Of Honor out of a Leo C Inman mare, Jetta C Leo, nicknamed "Cleo," was the other really great barrel horse that I've owned in recent years. He ranks in the same class with Cowhand Breeze. He was a 1992 futurity horse, winning over $48,000 that year. His largest paycheck, $22,453, came from the Reserve Championship of the World Championship Barrel Futurity. He picked up $12,701 that year for finishing fourth in the Old Fort Days Futurity. He also placed at the Texas Finest, Pineywoods, The Texas Barrel Race, San Antonio and Lazy E Futurities.

He won over $17,000 in his Derby year, by winning the Reserve Championship of the Lazy E Derby and Elite Barrel Derby. He picked up a $7,521 paycheck for finishing fourth at the Old Fort Days Derby, and $5,140 for fifth at the World Championship Barrel Derby. He was a great open barrel horse, with a big heart. Despite chronic feet problems, I won well over $125,000 on this great gelding before his untimely death in 1996.

Glossary

acupressure: the art of utilizing stimulation on pressure points to treat an animal.
acupuncture: the insertion of fine needles into specific, predetermined points on the surface of the body in order to regulate bodily functions.
all-out runs: running as fast as the horse will go.
alley-sour horse: a horse that will not come in the gate or entrance to the arena.
arc position: flexing a horse's body.
bad-minded horse: a horse that does not accept correction well and resists training.
balance rein: the outside rein. If it is picked up lightly, it can keep balance in the horse.
barrel futurity: a barrel race for horses from three to five years of age, depending on the rules of the particular futurity.
bars of the mouth: a portion of the horse's lower jaw which is devoid of teeth; the space between the wolf teeth (long, pointed teeth) and molars.
bell boots: protective boots that go over the horse's hoof and protect his coronary band.
big-hearted horse: a horse with exceptional desire to do well; a horse which accepts correction and wants to please his rider.
blister beetle: a type of beetle which invades fields of alfalfa hay and is toxic when ingested by horses.
blowing a barrel: coming out of a barrel too wide.
body signal: signaling a horse to go forward or slow down by the movement of a rider's body.
body throttle: body english; leaning forward to make a horse move forward and sitting back in the saddle to cue a horse to slow down or stop.
bonded: an established trust and a clear line of communication between a colt and a human.
breast collar: a strap that goes across a horse's chest and holds the saddle in place.
breeding: the pedigree or parentage of a horse.
broke horse: a horse that can be plow reined around and is gentle to ride.
bumping the bit: pulling lightly on the bit and then releasing to slow the horse down.
bunchy-muscled horse: a horse that is tight-made with muscles that are bunchy and protruding.
calf-kneed horse: a horse that is back in the knees.
cannon bone: the section of a horse's leg between his knee (in front) or hock (in back) and his pastern.
cantle: the rise behind the saddle seat.
carpal joints: knee joints.
cavesson: leather noseband which helps to keep a horse's mouth from opening.
chain-gag bit: a bit that works on the corners of the mouth; a bit used to regenerate a horse's mouth.
chargey horse: a horse that wants to run.
chiropractic: the science and art of utilizing the inherent recuperative powers of the body and the relationship between the nervous system and the spinal column, in the restoration and maintenance of health.
choppy stride: A horse with a really short stride which is hard to ride.

cinch: the strap around the horse's belly which attaches the saddle to the horse's back.
club-footed: a conformation flaw where the wall of a horse's foot is too steep, at a 60 degree angle or more; also called mule-footed.
combination bit: a bit where the mouth piece and the nose piece work simultaneously; a combination of a hackamore and a bit.
conditioning: an equine exercise program designed to develop musculature and wind power.
conformation: the general shape and size of a horse; the way in which a horse's body is put together.
control rein: the inside rein.
controlling a horse: directing a horse to use his abilities more effectively; maintaining command of the horse without dominating him.
cooling a horse out: walking a horse out until he is breathing normally and is mentally level.
correction bit: a bit with a solid bar; a tough bit with a port which usually creates a stiffer-turning horse.
cow-hocked: hocks that turn in and point toward one another instead of straight ahead, causing the horses toes to point out.
cowboy keg shoe: a regular keg horseshoe which does not have a rim.
crested neck: a thickening of the neck of the horse where the mane is.
cross leads: when a horse is on the correct lead in front but on the wrong lead behind.
curb chain: the chain that goes from one side of the bit to the other, placed under the horse's chin.
curb strap: a leather or nylon strap that attaches to the cheek pieces of a bit and works with the bit to exert leverage on the horse's mouth.
deworming: the use of a chemical dewormers to eliminate worms in the horse. Should be done after a fecal test has been run on the horse to determine that parasites are present.
disposition: a horse's attitude toward his handlers and other horses.
dog tracking: a horse's front legs which track wider apart than his back legs.
dog-legged horse: a horse's hind legs which sit way under him; also called a sickle-hocked horse.
double-hipped horse: a horse that has two "humps" of muscle at the top of his hip.
double-muscled horse: a horse with a crease down his rear end and down each side.
down on the front end: a horse shifting his weight to his front end while turning a barrel.
EPG: eggs per gram.
easy stop: a type of head gear that works under the horse's chin, on the jaw bone.
educated horse: a horse that picks up his leads, stops, moves his shoulders and responds to leg and rein cues.
electric eye: an electronic eye that starts the timer when the horse passes through the beam, and stops it when the horse breaks the beam again.
electrolytes: granules, paste or fluid put in a horse's feed or water that helps him stay hydrated.

elevation of the hock: a horse which moves with excessively high movement of the hock.
encephalomyelitis: sleeping sickness in a horse; an equine neurological disease.
estrus: the time when a mare comes into season and is ready to be bred.
feeling a horse's mouth: using the wrist and little finger lightly on the reins to just "feel" the horse's mouth.
finished horse: a barrel horse that can be "put on his honor" and will run a barrel pattern without much assistance from the rider.
flagging: use of a flag to time a horse as the horse crosses the starting line going to the first barrel and again as he crosses the line prior to leaving the arena.
flat-turning horse: a horse that goes to the barrel, disengages his rear end just a little, but just enough to keep himself self-propelled forward, yet never getting out of position to where he swings his butt.
flexion: the ability of a limber, well-conditioned horse to bend and turn his neck and body.
floating teeth: equine dental work where the sharp points of the horse's teeth, usually molars, are filed down to prevent the horse from injuring himself.
fluid motion: motion of a horse that is supple and soft.
foal: a horse (male or female) prior to weaning.
four-wheel drive turn: a style of turn where a horse runs to the barrel, collects and rates just enough to totally control the turn and uses all four legs, equally pulling with both front ones and pushing with both back ones separately.
free-choice feeding: having feed available at all times for a horse to eat.
free-up a horse: encouraging a horse to run.
free-wheeling gag bit: a dee-ring gag bit on a rope.
front end: the part of a horse's anatomy that includes the front legs, chest, shoulders and head of a horse.
front-end turn: a style of turn where a horse runs up to a barrel, keeps his backbone extremely straight and stiff and three-quarters of the way around the barrel, he drops the front end, moves his butt over, but only until it is in alignment to go straight to the pocket for the next barrel.
gaskin muscle: the large, inside, hind-leg muscle.
gelding: a castrated male horse.
german martingale: a piece of equipment that draws a horse's head down, yet has a direct-rein effect.
girth: the circumference of the horse's body measured from behind the withers and around the horse's barrel; the strap around the horse's belly which attaches the saddle to the horse's back—also called a cinch.
give you his head: a horse willingly performing a rider's instructions, via the reins, without resistance.
good muscle tone: muscles that are smooth and supple, with no visible or palpable muscle spasms or soreness.
good-minded horse: a horse that takes correction well and desires to please the trainer.
green: a horse that is barely broke.
hackamore: a type of bridle that has no bit.
hand: a standard measure for the height of a horse, equal to four inches.
hands on: having contact with and control of the horse.
hard pan: hard ground, usually below fluffy ground, in an arena which can cause the horse to go through the loose dirt on top and slip on the hard pan.

haulers: barrel racers who haul down the road to attend barrel races.
headstall: a leather strap that goes over the top of the horse's head to secure the bit.
heavy in the face: phrase used to describe a horse that is pulling on the reins against the rider's hands.
heavy on the front end: the act of a horse during a turn that causes most of his weight to be placed on his front end.
hill work: loping large circles on the side of a hill.
hobbles: a piece of equipment that goes around each of a horse's front legs to restrain him; used teach a horse manners, discipline and patience.
horsemanship: a rider's understanding of a horse; a sixth sense of what makes a horse "tick."
hot blood: a horse with running blood in its pedigree.
hot horse: an overly excited horse.
hot nail: a nail inside the white line of the hoof wall.
hot spot: a highly stressful place for a horse.
hyper horse: a highly nervous horse.
imprinting: desensitizing a colt after birth.
incisors: front teeth.
influenza: the most common viral cause of respiratory disease in a horse.
interdental spaces: the space between the front (incisors) and back (molars) teeth.
inside leg: the leg closest to the barrel in a turn.
inside leg pressure: pressure of the rider's leg against the horse on the side to which the horse is moving or turning.
jockey position: sitting forward in the saddle, urging the horse to run.
lead: a horse is said to be on the right or left lead depending on which front leg is leading in a lope — the lead foreleg strides forward farther than does the opposite foreleg.
leg cut in the swells: a saddle that has a cut-out in the swells so a rider's leg can fit close to the swell.
lifter: a bit with equal purchase and shank.
lizard-gutted horse: a horse with a shallow heart girth similar to a greyhound.
long stride: the act of a horse extending his back hoof print over his front hoof print.
long trot: a fast, extended trot.
maternal grandsire: grandsire on the dam's side.
mind's eye: a rider's sense of knowing exactly where a particular part of a horse is in any given moment.
molars: the back teeth of a horse.
mule-footed horse: see *club footed*.
mutton-withered: rounded withers.
open knees: a colt's front knees which have not developed enough.
outside gaskin muscle: the outside, hind-leg muscle of a horse.
outside leg pressure: pressure of the rider's leg against the horse on the side opposite the turn.
over flex: a horse that is over bent and too supple.
over in the knees: a conformation flaw where the front knees of a horse protrude beyond the cannon bones and ankle.
over-and-under: a rope that is locked in place on the saddle horn and used as a whip when encouraging a horse to run.
overbite: the projection of the upper teeth over the lower teeth; also called parrot-mouthed.
overreaching: striking the heel of the forefoot with the toe of the hind foot.

over-turning a barrel: continuing to turn a barrel even though it would ordinarily be time to go forward to the next barrel or run home.

oxbow stirrup: a round stirrup.

packed-cell volume: the blood count of a horse.

parrot-mouthed: an overbite where the top teeth project over the bottom teeth.

paternal grandsire: grandsire on the sire's side.

pedigree: a horse's parentage or bloodlines.

pig-eyed: a horse with small eyes.

pigeon-toed: a conformation flaw where a horse's toes point toward each other.

pin-eared: ears which are set close together on the horse's head.

plow rein: using direct rein to control the horse.

pocket: the space between the horse and the barrel.

poll: the area of the horse's head located between the ears which joins the skull with the spinal cord of the neck.

polo wrap: a rolled leg wrap used to wrap legs.

popping grain: a method of making grain highly digestible.

probiotic: fermented products which enhance the beneficial bacteria in the stomach.

prospect: a horse which, by its breeding, conformation and disposition, is potentially suited for a particular event.

purchase: the portion of the bit between the headstall ring and the mouthpiece.

putting a horse on his honor: allowing a horse to perform certain actions on his own without reining or rating him manually.

rabies: the neurological disease hydrophobia, caused by a virus which is transmitted by a wild animal bite.

rate: the degree to which you slow down or control the forward motion before and as you turn a barrel.

rear end turn: a style of turn where a horse turns on his rear end by setting and rolling back over his hocks.

releasing a horse: releasing pressure on the bit.

reverse arc: loping a circle in one direction and holding the horse's head in the opposite direction.

rhinopneumonitis: one of the most common forms of respiratory disease, particularly dangerous for pregnant mares.

riding a horse "into the bridle": putting slight pressure on the reins and urging the horse forward, which collects his body.

roping rein: a single rein going from one shank to the other, which prevents riders from dropping reins when switching hands.

roping stirrup: a heavy, wide stirrup used by ropers.

running through the bridle: the act of a horse having no regard for the bit.

sacking out: rubbing the horse's entire body with a cloth, blanket, sack, to desensitize a young horse.

saddle swells: the raised, front section of a saddle.

seasoned: a horse that has been hauled away from home and become accustomed to being in strange surroundings.

seesawing: pulling on one rein at a time, which works on one side of the horse's mouth at a time and keeps a horse soft in the body.

shank: the long cheek pieces of a bit.

short loping: slow loping.

short-strided: a horse's stride where a shorter distance is covered in one full cycle of limb motion.

shoulder twitch: a restraining technique; grabbing skin on a horse's shoulder, rolling it up and pulling it away from the horse's body.

sickle-hocked: weak hocks which are bent in the shape of a "sickle," resulting in cannon bones which are set at an angle, instead of perpendicular to the ground.

sidepull: a bitless bridle with a rope noseband over the nose and reins attached to each side.

snaffle bit: a jointed bit with o-ring or dee-ring cheekpieces, which exerts pressure on the corners of a horse's mouth.

solid horse: a barrel horse that can be "put on his honor" and will run a barrel pattern without much assistance from the rider.

speed control: the ability to adjust speed at a gallop through the use of cues.

splay-footed: a conformation flaw where a horse's feet point to the outside rather than straight forward.

splint boots: boots which cover the horse's front cannon bones to protect the splint bones from injury.

splints: a bump on the cannon bone of the leg caused by calcification from an injury to the splint bone.

strangles: streptococcus equi infection.

stride: a full cycle of limb motion; the distance covered by one foot when in motion.

strong breeder: a stallion which puts more of his traits in the offspring than the mare does.

supple: to soften or make flexible the horse's body parts through various exercises.

supplements: vitamins and minerals given in addition to a horse's regular feed.

tetanus: clostridium botulinum; a type of bacteria that invades open wounds or punctures and can cause a potentially life-threatening disease.

three-piece smooth snaffle: a three-piece snaffle bit with a smooth mouthpiece.

three-piece twisted snaffle: a three-piece snaffle bit with a twisted mouthpiece.

throttles down: the act of a horse slowing down before turning a barrel.

thrush: a degenerative condition of the frog characterized by a black discharge with a bad odor.

toe in: see ***pigeon-toed***.

toe-out: see ***splay-footed***.

tuning a horse: making a practice run.

twisted wire: a mouthpiece, usually jointed, made of twisted wire.

two-piece smooth snaffle: a snaffle bit with a smooth mouthpiece that breaks in two pieces.

two-piece twisted snaffle: a snaffle bit with a twisted wire mouthpiece that breaks into two places.

under turning a barrel: starting to run home or to the next barrel before the barrel is completely turned.

using a lot of flank: pressing a leg behind the flank cinch to help a horse get in position by bringing the hind leg and rear end under him.

vital approach area: the spot on the ground just before the barrel.

water founder: lameness caused by a horse drinking too much water when he is hot or stressed.

windpuffs: soft pockets of fluid caused by trauma in the fetlock joint, knee, or the pastern.

wolf teeth: small teeth immediately before the molars.

yearling: horses of either sex from Jan. 1 of the year after they were foaled until Jan. 1 of the next year, at which time they are considered two-year-olds.

Association Addresses

American Association of Equine Practitioners
4075 Iron Works Pike
Lexington, KY 40511-8434
Tel: (606) 233-0147 Fax: (606) 233-1968

American Farrier Association
4059 Iron Works Pike
Lexington, KY 40511-8434
Tel: (606) 233-7411 Fax: (606) 231-7862

American Horse Council
1700 K Street N.W., Suite 300
Washington, D.C. 20006
Tel: (202) 296-4031 Fax: (202) 296-1970

American Veterinary Chiropractic Association
623 Main Street
Hillsdale, IL 61257
Tel: (309) 658-2920

American Veterinary Medical Association
1931 N. Meacham Rd., #100
Schaumburg, IL 60173-4360
Tel: (708) 925-8070 Fax: (708) 825-1329

Barrel Futurities of America
Rte. 7, Box 395
Springdale, AR 72764
Tel: (501) 756-3107 Fax: (501) 756-9528

International Association of Dental Technicians
P.O. Box 6095
Wilmington, DE 19804-6095
Tel: (302) 892-9215 Fax: (302) 892-9999

International Veterinary Acupuncture Society
P.O. Box 1478
Longmont, CO 80502
Tel: (303) 682-1167 email: ivaoffice@aol.com

International Professional Rodeo Association
P.O. Box 83377
Oklahoma City, OK 73148
Tel: (405) 235-6540 Fax: (405) 235-6577

National Barrel Horse Association
P.O. Box 1988
Augusta, GA 30903-1988
Tel: (706) 722-7223 Fax: (706) 722-9575

National High School Rodeo Association
11178 N. Huron, Suite 7
Denver, CO 80234
Tel: (303) 452-0820 or 0912

National Intercollegiate Rodeo Association
1815 Portland Ave. #3
Walla Walla, WA 99362
Tel: (509) 529-4402 or 4403

National Little Britches Rodeo Association
1045 W. Rio Grande
Colorado Springs, CO 80906
Tel: (800) 763-3694

Women's Professional Rodeo Association
1235 Lake Plaza Drive #134
Colorado Springs, CO 80906
Tel: (719) 596-0900 Fax: (719) 576-1386

Index

Acupressure 169
Acupuncture 168
Adobe Joe King 21
Adonna 21
Alley-sour horses 154
American Association of Equine Practitioners 167
American Paint Horse Association 193
American Quarter Horse Association 193
American Veterinary Chiropractic Association 168
American Veterinary Medical Association 167
Astrodome 147

Babe Grande 21
Back boots 191–192
Band Of Honor 24
Bats 193
Beduino 23
Bell boots 191–192
Bennie's Big Red 23
Bert 21
Big circles 118–121
Billy Bars Bug 13, 196–197
Bits 184–189
   combination bit 186
   correction bit 189
   easy stop 189
   free-wheeling gag bit 188
   shank length 185–186
   snaffle 73–74
   solid mouthpiece bits 188–189
   three piece smooth snaffle 185
   three piece twisted snaffle 185
   two piece smooth snaffle 185
   two piece twisted snaffle 185
Blister beetles 171
Blondy Plaudit 21
Bloodlines 18–25
Bob's Folly 22
Boco Grande IV 22
Body throttle 72, 92–93
Bojingles 21
Bowden, Shy-Anne 197
Bozo 36
Breaking 66–73
Breast collars 183
Breeding 18–27
   age of broodmare 24–25
   bloodlines 18–25
   selecting a broodmare 24–27
   selecting a stallion 24–27
Broodmare
   age of 24–25
   selecting for breeding 24–27
Bugs Alive In 75 23
Bunny Bid 197
Burke, Laird 167
Burnt 22
Buying a barrel horse. See Selecting a barrel horse

Carey, Judy 197

Carter, John 194
Caulks 177–178
Cervi, Sherry 18
Cervi, Sherry Potter 199
Chief Breeze 194
Chiropractic 168
Cinches 182–183
Colborn, Everett E. 11
Combination bit 186
Communicating with a horse 54–55
Conditioning 130–135
   cooling out 135
   hill work 133
   stretching 134
   warming up 134–135
Conformation 28–30, 36–48
   back 44–46
   back legs 47–48
   buck-kneed 43
   calf-kneed 43
   cannon bone 42–43
   coronary band 48
   cow-hocked 47
   dog-legged 47
   ears 39–40
   eyes 39
   front end 41–43
   front leg alignment 43
   gaskins 47
   head 39–40
   heart girth 44
   hindquarters 46
   hocks 47–48
   hooves 48
   knees 42, 43
   mouth 39
   mule-footed 48
   neck 40
   nostrils 39
   pastern 44
   pigeon-toed 43
   shoulder 41
   sickle-hocked 47
   size 49
   splay-footed 43
   toe-in 43
   toe-out 43
   underline 44–46
   withers 44
Cooling out 135
Corks 177–178
Correction bit 189
Cowhand Breeze 11, 12, 194–196
Cowhand Joe 194
Cowhand Lady 194
Cues 90–93
   body throttle 92–93
   hand 90–92
   leg 92

Danish 14, 198–199
Dash For Cash 23
Day, Jeanna 21
Decker, Jo 12
Decker, Tater 12
Dental care 169–170

floating teeth 169–170
wolf teeth 169
Devilish Honor 199
Deworming 165
Disposition. See also Mind of a horse
   geldings 55
   mares 55–56
   stallions 56–57
   stomping 52
   tail popping 52
   teeth grinding 52
Doc Bar 24
Dorrance, Tom 61, 66
Drill Bar 198
Drills. See Exercises
Duck Dance 23

Easy Honor Jet 24
Easy Jet 23
Easy Max 23
Easy stop 189
Easy Thistle 23
Electrolytes 173
Encephalomyelitis 160, 162
Equipment. See Tack
Excuse 21
Exercises 116–127
   big circles 118–121
   four-cornered circles 126–127
   freeing up a horse 129
   half-arounds 125
   lengthening stride 97–99
   loops 121–123
   pasture loops 127–129
   spirals 125–126
   sprinting 102
   360s 82

Fairbert 13
Fast Jet 23
Feeding. See Nutrition
First barrel
   running past the barrel 154–155
Flagging 149–150
Flaming Jet 23
Flexion 49
Flicka Bob 22
Flicka Grande 22
Flit Bar 21
Floating teeth 169–170
Flying Bob 21
Fols Blue Six 23
Fols Native 23, 198
Four-cornered circles 126–127
Free-wheeling gag bit 188
French Flash Hawk 18
Futurity 115

Galls 166
Game Dame 12
Gate-sour horses 154
Geldings 55
Gender 55–57
Gills Bay Boy 22
Gill's Sonny Boy 23

Go Man Go 23
Goostree, Carol 22
Grady County 13
Grain 171–172
Ground condition 148, 150–151

Hackamore 186–188
Half-arounds 125
Harmon Baker 22
Hauling 144–151
    futurities 115
    ground condition 148, 150–151
    tips 148–149
    trailering 144–146
Hay 170–171
Health care 63, 160–169
    acupressure 169
    acupuncture 168
    blood work 166
    chiropractic 168
    cuts 166
    deworming 165
    encephalomyelitis 162
    galls 166
    influenza 162
    inoculations 160–162
    light skin 165–166
    massage 169
    rabies 162
    rhinopneumonitis 162
    soreness 166
    splints 179
    streptococcus equi infection 162
    swelling 162–163
    tetanus 162
    veterinarians 166–167
    windgalls 179
    windpuffs 179
    wrapping legs 163–165
Hobbles 67–68
Hocks
    soreness 178–179
Hoof care 174–179
    conditioners 174–176
    shoeing 177–178
    thrush 176
Hunt, Ray 66
Hyperactive horses 152

Imprinting 58–61
Influenza 160, 162
Inoculations 160–162
    encephalomyelitis 160, 162
    influenza 160, 162
    rabies 160, 162
    rhinopneumonitis 160, 162
    streptococcus equi infection 160, 162
    tetanus 160, 162
International Veterinary Acupuncture Society 168

James, Charmayne 18, 22, 50
Jerrys Bug 196
Jerry's Bug 13
Jet Deck 15, 23, 24, 25, 197
Jet Of Honor 14, 15, 23, 27, 199
Jet OJ 23
Jet's Pay Day 23
Jetta C Leo 14, 15, 24, 199
Jetta Jay 13, 197
Joak 21
Joe Reed 21
Joe Reed II 21

Johnny Boone 23
Johnson, Sherry Combs 21
Josey, Martha 22

Keg shoes 177–178
King 21
King Breeze 194
King Damon 21
King P-234 25
King York 21
Kirby, Connie Combs 21, 57

Lady's Bug Moon 13
Lameness 178–179
    sore hocks 178–179, 179
    windgalls 179
    windpuffs 179
Leads 77–79
Leatherman, Robyn 13, 196
Lee Bar 198
Leo 21, 23, 24, 25
Leo C Inman 14, 199
Leo's Question 21
Lightning Bar 24
Long trotting 132
Loomis, Joyce Burk 21
Loops 121–123

Major Movement 14, 197–198
Man O War Leo 21
Manual, Loretta 12, 194
Mares 55–56
Marion Breeze 13
Martingale 190–191
Massage 169
Maudie Leo 22
Maudies Joak 21, 57
Mayo, Jane 21
McLain, Shelly 18
Mechanical hackamore 186–188
Mehrens, Don 194
Mental preparation 136–143
Merganser 23
Mesquite, Texas 147
Mind of a horse 50–57. *See also* Disposition
    communicating with 54–55
    determining 52–54
    gender 55–57
Miss Elegant Bar 13
Monroe, Jimmie Gibbs 21
Mr Honor Bound 24

Nutrition 166, 170–173
    electrolytes 173
    grain 171–172
    hay 170–171
    probiotics 172
    supplements 172
    water 172–173

Oklahoma Star 21
Oklahoma Star Jr 21
Old Joe 22
On The Money Red 23
Orange Bug 13, 197
Osage Bob 21
Over-and-under 193
Overreaching 178

Packin Sixes 23
Pas Reb Pas 196
Pasture loops 127–129
Peter McCue 22

Peterson, Kristie 18, 36
Petska, Gail 21
Phillips, Becky 169
Plaudit 21
Pocket 85–87, 107
Poco Bueno 21
Poco Excuse 21
Poco Leo's Comedy 13
Poco Soto 21
Practicing 146–148
Probiotics 172

Quarter Horse News 14, 23
Quirts 193

Rabies 160, 162
Raise Your Glass 23
Rally Racer 197
Ramblin Rally 13
Rambling Rally 197
Rare Jet 23
Rate 34, 85, 103–107
    setting up too much 158–159
    sore hocks 178–179
Red Buck 22
Reins 189–190
Rene Dan Jet 24
Rhinopneumonitis 160, 162
Rim shoes 177–178
Robin Flit Bar 21
Robin Hood Price 21
Rudy Val 24

Sacking out 69
Saddle pads 182
Saddles 180–182
Sanders, Janice 197
Saunders, Jack 11
Saunders, Janice 13
Scamper 18, 22, 50
Scooter Waggoner 21
Second barrel
    running past the barrel 154–155
    turning before 155
Secretariat 23
Selecting a barrel horse 28–35
    conformation 28–30
    for novice riders 28
    judging mental attitude 32
    judging physical ability 28–31
    price 34
    style of turning barrels 33–34
    vet checks 34–35
Sexy Classy Moon 24
Shoeing 177–178
    caulks 177–178
    corks 177–178
    keg shoes 177–178
    rim shoes 177–178
Shoot Yeah 23
Showdown 21
Showing. *See* Hauling
Sir Double Delight 18
Six Fols 23
Sleeping sickness. *See* Encephalomyelitis
Snaffle 73–74
    combination bit 186
    two piece smooth snaffle 185
Snaffles
    three piece smooth snaffle 185
    three piece twisted snaffle 185
    two piece twisted snaffle 185

Sonny Bit O'Both 22
Sonny Gill 23
Sooner Satin 21
Soreness 166
   hocks 178–179
Special Effort 23
Speed
   first barrel 103
   increasing 100–114
     second barrel 108–110
     sprinting 102
     third barrel 110–115
Spirals 125–126
Splint boots 179, 191–192
Splints 179
Spoiled horses 152–154
Sprinting 102, 132
Spurs 192–193
Stallions
   disposition 56–57
   hauling 57
   selecting for breeding 24–27
Star Plaudit 21
Starting colts 64–79
   bitting 73–74
   breaking 66–73
   duration of training 77
   first saddling 68–70
   hobbling 67–68
   leads 77–79
   mental maturity 66
   mounting 70–73
   physical maturity 64–66
   sacking out 69
   stopping 75
   teaching cues 74–75
Stomping 52
Strangles. *See* Streptococcus equi infection
Streakin Six 23
Streptococcus equi infection 160, 162
Stress 89–90
Stretching 134
Stride 49, 178
   lengthening 97–99
Sugar Bars 21
Sunshine Plaudit 21
Supplements 172
Suppling 116–118. *See also* Conditioning
Swelling 162–163

Tack 180–191
   back boots 191–192
   bats 193
   bell boots 191–192
   bits 184–189
   breast collars 183
   cinches 182–183
   combination bit 186
   correction bit 189
   easy stop 189
   free-wheeling gag bit 188
   martingale 190–191
   mechanical hackamore 186–188
   over-and-under 193
   quirts 193
   reins 189–190
   saddle pads 182
   saddles 180–182
   solid mouthpiece bits 188–189
   splint boots 179, 191–192
   spurs 192–193

three piece smooth snaffle 185
three-piece twisted snaffle 185
tiedown 190
two piece smooth snaffle 185
two piece twisted snaffle 185
Tail popping 52
Te N'Te 23
Teeth grinding 52
Tetanus 160, 162
The Ole Man 196, 197
Third barrel
   coming out of wide 155–156
Three Bars 21, 22, 23, 24, 198
Three Oh's 23
Three piece smooth snaffle 185
Three piece twisted snaffle 185
Thrush 176
Tiedown 190
Timers 149–150
Tiny Charger 23
Tommy Clegg 21
Tompkins, Harry 11
Tonto Bars Gill 23
Top Bar Bug 13, 196
Top Deck 23
Top Moon 23
Trailering 144–146
   feeding in trailer 144–146
   trailer preparation 146
   wrapping legs 144
Training
   advancing 87–90
   alley-sour horses 154
   big circles 118–121
   bitting the colt 73–74
   breaking 66–73
   coming out of third barrel wide 155–156
   conditioning 130–135
   correcting the horse 97
   cues 90–93
   dropping front end 158
   duration for colts 77
   exercises 116–127
   first barrel 103
   first saddling 68–70
   four-cornered circles 126–127
   four-wheel drive turn 94–95
   freeing up a horse 129
   front-end turn 96–97
   gate-sour horses 154
   half-arounds 125
   hand cues 90–92
   hitting barrels 156–158
   hobbling 67–68
   hyperactive horses 152
   imprinting 58–61
   increasing speed 100–114
   leads 77–79
   leaving the barrel 107–108
   leg cues 92
   lengthening stride 97–99
   loops 121–123
   mental maturity 66
   mental preparation 136–143
   mounting 70–73
   pasture loops 127–129
   physical maturity 64–66
   pocket 85–87, 107
   problems 152–158
   rate 85, 103–107
   rear-end turn 96
   running past the barrel 154–155

sacking out 69
second barrel 108–110
setting up too much 158–159
spirals 125–126
spoiled horses 152–154
sprinting 102
starting colts 64–79
starting on barrels 80–98
stopping 75
stress 89–90
teaching cues 74–75
third barrel 110–115
360s 82
turning barrels 93–97
turning before second barrel 155
yearlings 58–63
Tuno 21
Tuno's Lady 21
Turning barrels 93–97, 158–159
   coming out of third barrel wide 155–156
   dropping front end 158
   first barrel 103
   four-wheel drive turn 94–95
   front-end turn 96–97
   hitting barrel 156–158
   leaving the barrel 107–108
   rear-end turn 96
   second barrel 108–110
   style of turns 33–34
   third barrel 110–115
   turning before second barrel 155
Two piece smooth snaffle 185
Two piece twisted snaffle 185

Up N Truckle 23

Vernon, Texas 147
Veterinarians 166–167
V's Sandy 21

War Leo 21
Warming up 134–135
Water 172–173
Whittle, Jerry 13
Wimpy 21
Windgalls 179
Windpuffs 179
Wolf teeth 169
Wrapping legs 144, 163–165

Yearlings 58–61, 63
   training 58–63
Youree, Dale 13
Youree, Florence 13